A Naturalist's Guide

BI
OF CHINA

Southeast China including Shanghai

Liu Yang, Yong Ding Li and Yu Yat-tung

JOHN BEAUFOY PUBLISHING

First published in the United Kingdom in 2014 by John Beaufoy Publishing Ltd
11 Blenheim Court, 316 Woodstock Road, Oxford OX2 7NS, England
www.johnbeaufoy.com

10 9 8 7 6 5 4 3 2 1

Photo Credits
Front cover: *top, from left to right:* Fairy Pitta (Bjorn Olesen), Cabot's Tragopan (Abdelhamid Bizid), Short-tailed Parrotbill (Tang Jun); *bottom, left to right:* Chinese Crested Tern (Abdelhamid Bizid); Blue-crowned Laughingthrush (Forrest Tong).
Title page: Spoon-billed Sandpiper © Peter and Michelle Wong
Contents page: Chinese Barbet © John and Jemi Holmes
Christian Artuso 17t; Mikael Bauer 104t; Abdelhamid Bizid 19bl, 20t, 20b, 22t, 24b, 25b, 26t, 27br, 33bl, 35t, 38bl, 39b, 40b, 42t, 44b, 54b, 56t, 57t, 57br, 59t, 59b, 61b, 65b, 67tl, 75b, 76tr, 77b, 78tl, 80br, 83t, 84bl, 86t, 87t, 87b, 88t, 89tr, 92b, 95t, 96b, 100b, 108br, 109t, 109b, 111t, 113t, 114b, 115tl, 117b, 121b, 123b, 125t, 125b, 126tr, 126bl, 127t, 127b, 128t, 129b, 130b, 131t, 131b, 132b, 134b, 135tl, 135b, 136t, 140b, 141b, 142tl, 142b, 143tl, 144bl, 145tr, 149b, 150tr, 151bl, 152tl, 152b, 153t, 153b, 155t, 155bl; Ah Kei looking@Nature 26b, 37t, 38t, 48t, 50b, 68br, 70t, 72t, 132t, 137t, 137b, 141t, 152tr; Sam Chan 93t; Chen Qing 21t; Cheng Heng Yee 30t, 34b, 41t, 97b; Frankie Cheong 50t, 139tr; Chung Weng Kin 31b; James Eaton 18br, 46b, 99b, 105b, 151t; Forrest Fong 116t; Sundev Gombobaatar 21b, 23t, 23b, 44t, 45t, 48br, 66bl, 155br; Martin Hale 60t, 106t; Hao Xiang 74tr; John and Jemi Holmes 33tr, 46t, 49tl, 58t, 66br, 67tr, 69t, 70b, 74t, 81b, 82t, 89tl, 90b, 115b, 123t; Kinni Ho 16t, 36tl, 36b, 40t, 63b, 65t; Mike Kilburn 79br; Koel Ko 22b, 45b, 61t, 71t, 71b, 72b, 75tl, 95b, 96t, 106b, 122t, 124b, 147t, 148t; Le Manh Hung 28b; Lee Kam Cheong 74b, 149t; Lee Tiah Khee 15b, 54t, 62t, 68t, 98b, 103b; Lee Shunda 80bl; Jennifer Leung 55t, 140t; Low Choon How 38br; Lu Gang 17b, 113b; Daisy O'Neill 19br; Bjorn Olesen 73b; Stuart Price 91b; Sun Xiaoming 24t; Tan Gim Cheong 51t; Thiti Tan 79bl; Tang Jun 18t, 18bl, 94t, 116b, 119b, 120t; Myron Tay 35b, 42b, 49b, 57bl, 81t, 85b, 88b, 119t, 126tl, 128b, 133b, 143tr, 143b, 145tl; Ivan Tse 147b; Wallace Tse 101t; Tong Menxiu 31t, 43b, 45b, 53t, 76tl, 78b, 82b, 83b, 84br, 91tl, 93b, 100t, 104b, 107t, 107b, 110b, 111b, 114t, 115tr, 117t, 118t, 120b, 121t, 122b, 126br, 134t, 142tr, 148b; Wang Jiyi 39t; Jason Wong Wai Hang 68bl; Michelle and Peter Wong 16b, 25t, 33br, 34t, 36tr, 37b, 47tr, 49tr, 53b, 55b, 58b, 60b, 62b, 63t, 64t, 64b, 67b, 77t, 78tr, 79t, 84t, 85t, 86b, 90t, 91tr, 92t, 94b, 97t, 98t, 99t, 101b, 102t, 102b, 103t, 108t, 110t, 112t, 112b, 118b, 124t, 129t, 133t, 134tr, 138b, 139b, 145b, 146b, 150tl, 150b, 151bl, 154t, 154b; Francis Yap 27bl, 28t, 29tr, 29b, 30b, 32t, 32b, 43t, 47b, 48bl, 51b, 52t, 52b, 66t, 69b, 80t, 136b, 138t, 139tl, 144r, 144br; Yong Ding Li 27t, 29tl, 89b, 130t; Zhang Ming 56b, 73t, 146t; Zhang Yong 108bl; Zhao Jian 33tl, 41b, 47tl, 105t.

ISBN 978-1-909612-23-5

Edited and designed by Gulmohur

Printed and bound in Malaysia by Times Offset (M) Sdn. Bhd.

·Contents·

Introduction

Despite more than two millennia of settlement and a concentration of some of the world's largest metropolises, Southeast China is home to a remarkably diverse avifauna, including sought-after birds like Cabot's Tragopan, Blue-crowned Laughingthrush and the Critically Endangered Chinese Crested Tern. The wetlands of Southeast China sit along one of the Earth's most important migratory flyways and support hundreds of thousands of waterbirds, including a fine set of the world's most charismatic ducks (for example the Mandarin Duck). In spite of these draws, birdwatching has been slow to take off. For years, there was little information on birdwatching sites and bird distributions across this region. Language and limited access in remote areas also proved to be barriers. Fortunately, this is changing. Southeast China's birdwatching scene, originally fuelled by interest from overseas birdwatchers is now growing rapidly, while vastly improved infrastructure has opened up a plethora of previously inaccessible sites. Yet many areas remains poorly-known ornithologically, especially the rugged mountains of Hunan, Zhejiang and Hainan, thus presenting great opportunities for exciting discoveries to be made.

Using photographs only of birds in a wild state, this book introduces 280 species of the region's avifauna. While we aim to cover as representative a sample of Southeast China's birdlife, and illustrated subspecies from the region where possible, we have omitted a number of well-known Oriental and Palearctic species, especially many widespread waterbirds. In their place are species more closely associated with the region. In cases where these omitted species are similar to those described, they are briefly described for comparisons, and are highlighted in bold. Additionally, we also provide some practical information for birdwatchers planning trips to the region. We hope our work will spur interest in the avifauna of Southeast China among birdwatchers, nature-lovers and conservationists.

Geography

The geographical scope of this book covers an 820,000 sq km segment of China south of the Yangtze River, including six provinces (Hunan, Jiangxi, Zhejiang, Fujian, Guangdong and Hainan) and one municipality (Shanghai). Southeast China is characterized by the watersheds and floodplains of the middle-lower Yangtze and the Zhujiang (Pearl) Rivers. A number of major tributaries and shallow lakes drain into the Yangtze, including the two largest freshwater lakes in China - Poyang Lake (c. 4,125 sq km) and Dongting Lake (c. 4,000 sq km). The extent of both lakes fluctuates greatly, but is declining due to land reclamation, sedimentation and droughts. The most important rivers in the region are tributaries of the Yangtze, namely the Xiang, Zi, Li, Ruan, Gan and the Pearl River. The Pearl River is China's third longest river, with its mouth marked by a large bay separating the cities of Macau, Shenzhen and Hong Kong. Both the Yangtze and Pearl River basins are very fertile and have been extensively cultivated for millennia.

Southeast China is generally mountainous and features four major ranges. Three of these, the Wuling-Xuefeng Mountains in western Hunan, the Mufu-Jiuling-Luoxiao Mountains that straddle the border between Hunan and Jiangxi, and the Yandang-Wuyi-Daiyun Mountains, which stretch from Zhejiang/Jiangxi to Fujian run north-south. The

Nanling Mountains run east-west from eastern Guangxi and southern Hunan to northern Guangdong and southern Jiangxi. Some of the highest peaks in the region are Huanggangshan (2,161m) in the Wuyi Mountains, Nanfengmian (2,120m) in the Luoxiao Mountains, Hupingshan (2,099m) in the Xuefeng Mountains and Shikengkong (1,902m) in the Nanling Mountains. Peripheral to the mountains are low hills forming three major hill zones: Guangxi–Guangdong, Jiangnan and Zhejiang–Fujian.

Forested slopes of the Wuyi Mountains at the Jiangxi-Fujian border.

Located in the South China Sea, and separated from the Leizhou Peninsula by the 80km-wide Qiongzhou Straits, Hainan is the second largest island in China after Taiwan. Hainan's interior is rugged and is characterized by steep mountains rising to 1,840m at Wuzhishan, bounded by a narrow coastal plain.

CLIMATE

Southeast China is a transitional region that extends across the subtropical and tropical climatic zones. The subtropical zone that covers the northern half of the region is characterized by wet monsoons, as well as large variations in temperature and rainfall between winter and summer. Winters are relatively long, cold and dry due to cold winds blowing from northern China. Average temperatures in January range from 4 to 13°C, with much variation across the region. In the mountainous areas frost is common and snowfall occasionally occurs. Summer is characterized by high temperatures, with frequent and heavy rainfall. Average summer temperatures in July are approximately 25–31°C, while average annual rainfall varies between 1,000 and 1,500mm. Spring and autumn are short, with variable weather. In lower latitudes like southern Guangdong and Hainan, the climate is milder, but subject to the influence of tropical monsoons. While temperature variations between months are small and usually within a range of 10–15°C, variations in monthly rainfall are larger and are heavily influenced by typhoons that blow from the West Pacific. Average annual rainfall across this region varies from 1,500 to 2,000mm.

AVIAN BIOGEOGRAPHY OF SOUTHEAST CHINA

Southeast China's extensive mountains present great physical barriers and points of isolation to birds with poor dispersal ability. During the Pleistocene glacial periods, isolation of bird populations on mountains drove speciation. Many endemic species and subspecies, particularly of pheasant, babbler and laughingthrush in Southeast China are likely to have arisen this way after periodic isolation from related forms in the Eastern Himalayas, southwest China or Southeast Asia. Similarly, populations isolated on islands often follow divergent evolutionary trajectories. This is best exemplified by high levels of endemism at the species (four) and subspecies (c. 55) level on Hainan Island.

From a biogeographical perspective, mountains not only act as barriers, but also provide varied ecological opportunities for species diversification. During warmer interglacial periods some species may experience range expansions, followed by colonizations of habitats and niches along the elevational gradient. Altitudinal partitioning of closely related species, such as *Phylloscopus* and *Seicercus* warblers, demonstrates such evolutionary processes. On the other hand, the occurrence of species associated with the Himalayas (for example Brown Bullfinch and Spotted Laughingthrush) in some mountains of the region is likely the consequence of colonization during warmer interglacial periods.

Unlike Southeast Asia, Southeast China is not a hotspot for species discovery. Only one bird new to science has been described in the past 20 years (Hainan Leaf Warbler). However, phylogenetic insights from recent studies have elevated several endemic subspecies in the region to full species, especially taxa in the Grey-cheeked Fulvetta, Black-throated Laughingthrush, Rufous-necked Scimitar Babbler and Blyth's Leaf Warbler species-complexes. Despite this, further phylogenetic work is necessary to better understand the evolutionary origins of Southeast China's avifauna, particularly in relation to Taiwan, Indochina and the Eastern Himalayas.

Habitat types and bird communities

Southeast China's diverse habitats support slightly over 700 species, or roughly half of China's avifauna. Of this, about 380 species breed in the region (residents and summer visitors), with the remainder mostly passage or wintering migrants. Here we describe key habitats and their associated bird communities.

FORESTS

The climax vegetation across most of Southeast China is broadleaved evergreen and mixed forests dominated by oaks, chestnuts (*Quercus, Castanopsis* spp.) and other genera such as *Cinnamomum, Ilex* and *Phoebe*. At higher elevations and with increasing

Broadleaved evergreen forests in the hills of Jiulianshan Nature Reserve.

Large stands of bamboo occur at the lower elevations of Wuyishan.

disturbance, deciduous genera and conifers become more prominent. Floristic diversity across Southeast China is high, and well-studied Dinghushan in Guangdong supports nearly 2,000 vascular plant species. Unfortunately, much of the region's forests have been cleared or disturbed, so remaining forests are mainly secondary, and dominated by fast-growing pines like *Pinus masonniana* and *Cunninghamia lanceolata*, as well as mostly planted stands of bamboo (*Phyllostachys pubescens*). Undisturbed forests are mostly confined to the upper slopes of high mountains from above 1,500m.

Bird communities in Southeast China's subtropical forests are essentially Sino-Himalayan, being characterized by *Phylloscopus* and *Seicercus* warblers, tits, flycatchers, fulvettas, yuhinas, laughingthrushes and bulbuls. Galliforms are well represented with nine species, of which three, Cabot's Tragopan, Elliot's Pheasant and White-necklaced Partridge, are endemic to the region. Many sites support three to six species of laughingthrush. Towards western Hunan, forest bird communities contain a number of species that are more typical of the Sichuan (Southwest China) uplands, and include birds such as Temminck's Tragopan, Rufous-gorgetted Flycatcher and Black-faced Laughingthrush. Many tropical families like trogons, bulbuls and barbets are also represented, but with less diversity compared to Southeast Asia.

The forests of Hainan and southernmost Guangdong are subject to wetter climates and support floristically distinct evergreen tropical monsoon forests. Lowland forests in Hainan, for instance, are dominated by Southeast Asian dipterocarp genera such as *Vatica* and *Hopea*. At higher elevations chestnuts, oaks and laurels increasingly dominate, together with tropical conifers like *Dacrydium*. Bird communities are richer and include many tropical species at the northern limits of their distribution, among them Silver-breasted Broadbill, Blue-bearded Bee-eater, Blue-rumped Pitta and Spot-necked Babbler.

Tropical monsoon forests at submontane elevations, Jianfengling, Hainan Island.

In Southeast China forests dominated by conifers are usually confined to heights above 1,500m. On the Wuyi Mountains for instance, Hemlock (*Tsuga chinensis*) dominated forests become increasingly ubiquitous from 1,400m. The highest elevations of the region's mountains are usually covered with shrubby vegetation, including bushes such as *Rhododendron, Stranvaesia* and various upland grasses. Bird communities in coniferous forests and alpine shrubland are impoverished and characterized by tits, fulvettas and *Phylloscopus* and *Hornornis* warblers.

Mangrove forests used to be extensive on the coast to as far north as Zhejiang, but have been extensively cleared. Small patches remain on Hainan, the Leizhou Peninsula and Pearl River delta, and in southern Fujian. Floristic diversity is relatively low, with fewer than 25 true mangrove species. Bird communities of mangroves are species poor, and characterized by habitat generalists like Common Tailorbird and Greater Coucal.

▪ HABITAT ▪

WETLANDS

Flood lakes are ubiquitous on the lower alluvial plains of the Yangtze River in north Hunan and Jiangxi. Two of the most important lake systems for birds are the Dongting and Poyang Lakes. During the dry season, lowered water levels expose vast expanses of flats, waterlogged grassland, marshes and *Phragmites*-dominated reed beds. These important wetlands support large wintering populations of waterfowl, cranes, storks, rails, gulls, shorebirds and passerines.

Freshwater wetlands in Poyang Lake support large numbers of waterfowl, and cranes

The Southeast China coast also holds significant stretches of coastal flats, reed beds and salt marshes. The mouths of the Min and Pearl Rivers support extensive mud and sand flats, with some of the best examples to be found at Deep Bay on the Hong Kong/Guangdong border. Rich in worms, crabs and bivalves, these flats attract large flocks of shorebirds, gulls and ducks. Coastal reed beds and salt marshes now occur in scattered patches, with the largest found on the Zhejiang-Shanghai coast, and Chongming Island in the Yangtze delta. Besides waterbirds, these habitats are also important to resident and wintering passerine assemblages. The invasion of Southeast China's coastal marshes by the Smooth Cord Grass (*Spartina alterniflora*) of North America is rapidly altering the structure of these wetlands, with implications for its bird communities. Furthermore, many of the natural wetlands along the coast have now been converted to prawn and fish ponds, which make them less attractive to most wintering waterbirds, although some ardeids and shorebirds do utilize these habitats.

AGRICULTURAL AREAS

Much of Southeast China, particularly the coastal plains of Guangdong, Fujian and Hainan, and the alluvial plains of Hunan, are now intensively farmed. Cultivation also extends to the lower slopes of many of the region's uplands. For example, parts of the Wuyi Mountains have long been cleared for tea and bamboo cultivation. Abandoned

farmland develops low, shrubby vegetation that may succeed to low-stature woodland. Breeding bird communities in these human-modified habitats are impoverished, and are represented by a few adaptable species of shrike and starling, Common Tailorbird and Common Magpie. Paddy fields and waterlogged farms, on the other hand, are analogous to wetland habitats and attract resident and migrant herons, rails, shorebirds and passerines.

Tea cultivation on the lower slopes of Wuyishan.

URBAN PARKLAND

There are thousands of manicured parkland patches across the cities and towns of the region. As heavily modified landscapes with low floral diversity, most parks support few but the most adaptable birds, including Red-whiskered and Light-vented Bulbuls, and Spotted Dove. However, many passage migrants like flycatchers, thrushes and warblers also utilize these patches of urban greenery to rest and feed in spring and autumn.

OFFSHORE ISLANDS

The coasts of Zhejiang, Fujian and Guangdong are dotted with numerous small island groups, many of which are rugged and uninhabited. All except the Matzu and Kinmen islands (Taiwan controlled) are administered by China. These islands support breeding terns, Black-tailed Gull, Pacific Reef and Chinese Egret, and the last colonies of the Chinese Crested Tern.

BIRD MIGRATION AND THE EAST ASIAN-AUSTRALASIAN FLYWAY

Southeast China lies within the East Asian-Australasian Flyway, a globally important flyway for migratory birds. The northeast extreme of this flyway is the subarctic tundra of eastern Siberia and Alaska, a region that supports millions of breeding ducks, geese and shorebirds in spring. South of the tundra region is a vast belt of forest steppe, coniferous and mixed broadleaved forests stretching across eastern Russia, northeast China, the Korean Peninsula and the Japanese Archipelago. This region supports more than 200 species of breeding landbird, including many raptors, cuckoos and passerines. As summer ends, many birds start to depart their breeding grounds for the warmer latitudes of southern China, Southeast Asia and Australia.

Depending on the rate of migration and various environmental factors, the first southbound migrants start arriving in Southeast China from midsummer. In Shanghai, shorebirds arrive as early as the start of July, while songbird migration tends to pick up towards the end of July, peaking in late August. Large waterbirds including gulls, geese and cranes may not arrive until November-December. Further south in Guangdong and Hainan, first arrival timings for most migrants usually lag by a number of weeks. For instance, Yellow-rumped Flycatchers start arriving by the last week of July in Shanghai, but not until early September in Guangdong. Similarly, Rufous-tailed Robins, which breed even further north, arrive in northern Zhejiang from October, reaching Guangdong from November.

The habitats of Southeast China form important 'connecting' and staging sites for migratory birds moving on to Southeast Asia and Australia. Many species that occur in the region are 'passage migrants' that stop to feed and rest before continuing on their migration to wintering destinations further south. Examples of passage migrants include Siberian Blue Robin and Arctic Warbler, both abundant songbirds that winter in Southeast Asia. A number of species occur as both passage migrants and winter visitors, especially waterbirds with wide wintering distributions (such as Eurasian Curlew).

Southeast China is also an important wintering region for a large number of birds, supporting hundreds of thousands of shorebirds and waterfowl, and including many of the most threatened waterbirds in the world. Nearly the entire population of the Critically Endangered Siberian Crane relies on the wetlands of the Poyang and Dongting Lakes as wintering habitat. While numerous passerines, including Japanese Robin and Tristram's Bunting, have most of their wintering ranges in Southeast China, there are a number of species that have much of, if not their entire breeding ranges in Southeast China. Among these are long-distance

Wintering shorebirds on the Minjiang Estuary in Fujian.

migrants like Brown-chested Jungle Flycatcher, and short-distance migrants like Silver Oriole, both of which mostly winter in Southeast Asia's rainforests.

BIRD CONSERVATION

According to the IUCN Red List, 38 globally threatened species are residents, breeding or winter visitors to Southeast China. Five are listed as Critically Endangered, ten as Endangered and 23 as Vulnerable, more than any European country. Supporting two-thirds of the region's threatened species (26), wetlands are among the most threatened habitats. Many are near heavily populated areas and are thus often lost to development or pollution, a situation best exemplified on Chongming Island's remaining wetlands, or subjected to disturbances like fishing, hunting and recreation. Fewer forest species are globally threatened (12), but many have narrow distributions. As a result of extensive historical deforestation and disturbance, remaining forests are heavily fragmented and many are isolated on mountains. Populations of some of the most threatened species like White-eared Night Heron and Silver Oriole are now confined to these fragments, the latter being further impacted by habitat loss in its Southeast Asian wintering grounds.

Apart from habitat loss and degradation, another major threat is the prevalent trapping of birds for food and the pet trade. Consumption of wildlife, including birds, is a major conservation issue in Southeast China. Many ducks, geese, shorebirds, pheasants and even small passerines are hunted for food, while species like Greater Coucal and most owls are believed to hold medicinal properties. Most notably, Yellow-breasted Bunting is considered a local delicacy and has suffered from large-scale trapping, with drastic population declines in recent years. An established tradition of keeping birds as pets has driven demand for species with melodious songs like Chinese Hwamei, Oriental Magpie Robin and Oriental Skylark, resulting in them being trapped in large numbers. While many of these popular songbirds appear common in the wild, due to the lack of monitoring the impacts on their populations remain unknown.

Laws that protect wild animals from hunting, like those that target illegal land development, have been promulgated, but enforcement remains inadequate. Low awareness of biodiversity and conservation among the public has limited the support given to governments and nature-reserve managements in tackling conservation issues. However, while birdwatching is still a relatively new concept, the situation is changing rapidly and there are now established birdwatching groups in many coastal cities. Presently, there are more than 30 active birdwatching societies in China, with some even making the first inroads into more proactive advocacy. Collaborative work in the form of the long-term China Coastal Waterbird Census, initiated in 2005, involves 150 volunteers from various stakeholder organizations across nine provinces and cities, and has provided considerable data to inform conservation activities. The Census recently received international recognition at the Ford Green Awards in 2012. Additionally, the BirdLife China Programme, which was jointly launched in 2005 by BirdLife International and the Hong Kong Bird Watching Society, has helped to establish new birdwatching clubs across Southeast China, besides initiating several species-specific conservation projects targeting Blue-crowned Laughingthrush, Chinese Crested Tern and Spoon-billed Sandpiper. While these developments offer optimism for nature conservation in Southeast China, an even greater challenge is for local governments across the region to balance the goals of rapid economic development with the active conservation efforts that the region so urgently needs.

Birdwatching in Southeast China

With the growing availability of organized tours, many birdwatchers are now able to enjoy the region's birdlife. Although access to some of the remote reserves is still limited, the many publicly accessible reserves and scenic sites allow birdwatchers to travel independently to see a good selection of the region's specialities. Given the seasonal occurrence of migrant birds, it is not possible to see all of the specialities within a two-week trip. However, a good selection of species may be found on a carefully planned trip in September–October, November–January or March–April. Those planning to see the specialities that breed in the region should also consider trips in April–May.

SHANGHAI AND OUTSKIRTS

Sited by the Yangtze delta, Shanghai's environs are among the best places to observe migratory birds. One of the most popular sites is **Nanhui Dongtan Wetland Reserve** (南汇东滩), a large area of reed beds, lakes and coastal flats south of the city. Key species here include Reed Parrotbill, Japanese Swamp Warbler, Chinese Penduline Tit and many waterbirds. The largest park in Shanghai, the well-wooded **Century Park** (世纪公园) provides an excellent introduction to common birds in the region, including a good selection of migrant passerines that stop over during the passage months. Access is on Line 2 of the Shanghai Metro. About 40km to the north of Shanghai is the **Chongming Dongtan Wetland Reserve** (崇明东滩) on Chongming Island, another excellent area for waterbirds and passerines. Over 300 species have been recorded in the woodland, reed beds and tidal flats here.

ZHEJIANG PROVINCE

Xixi National Wetland Park (西溪国家湿地公园) Located within Hangzhou, Zhejiang's provincial capital, the woodland and wetlands of Xixi offer an easily accessible introduction to the common waterbirds of the region, given its extensive network of trails. Many migratory thrushes and flycatchers are also possible in winter. The scenic **West Lake** (西湖) supports some waterbirds, and is especially known for the large flocks of wintering Mandarin Ducks. Alternatively, the well-wooded **Hangzhou Botanical Gardens** (杭州植物园) and **Zijingang campus of the Zhejiang University** (浙江大学紫金港校) are worth exploring and very popular with local birdwatchers.

Xiaoyangshan Island (小洋山岛) Home to the Yangshan deepwater port, Xiaoyangshan is one of the best sites in the region for observing eastern Palearctic migrants. The patches of scrub and woodland supports numerous warblers, flycatchers, thrushes and chats during the spring and autumn passage period, including sought-after migrants such as Japanese Paradise and Blue-and-white Flycatcher. The island can be accessed by the 32km Donghai bridge.

JIANGXI PROVINCE

Wuyuan County (婺源) The landscape of this northeastern county in Jiangxi is varied and hilly, and supports a mosaic of evergreen mixed forests, farmland and bamboo stands. It is probably the only site in the world where the highly localized Blue-crowned Laughingthrush can be regularly seen, best in spring/summer. The threatened Scaly-sided Merganser also regularly winters in Wuyuan's rivers in small numbers, as do large numbers of the attractive Mandarin Duck. Lodging is available in Wuyuan city.

Poyang Lake Wetlands (鄱阳湖湿地) The wetlands of the Poyang Lake system are among the most important inland sites for wintering waterbirds in China, supporting large numbers of wildfowl and cranes. This is also an important wintering site for the rapidly declining Baer's Pochard and enigmatic Swinhoe's Rail. Two frequently visited parts of the wetlands are the **Nanjishan** (南鹿山自然保护区) and **Poyang Hu (National) Nature Reserves** (鄱阳湖自然保护区). Access is from Nanchang, Jiangxi's provincial capital.

Wuyishan (National) Nature Reserve (武夷山自然保护区) Famed for their scenic beauty and tea cultivation, the forested Wuyi Mountains of east Jiangxi and west Fujian are home to the highest peak in the region, Mount Huanggang (2,161m). This reserve protects a long list of south Chinese specialties, particularly Cabot's Tragopan, White-necklaced Partridge and Short-tailed Parrotbill. Most of the region's montane forest birds can also be seen here. A permit is needed for access.

Jiulianshan (National) Nature Reserve (九莲山自然保护区) Straddling the rugged border between Jiangxi and Guangdong, Jiulianshan is best known for being one of the prime sites to see the White-eared Night Herons in China. The reserve's extensive evergreen forests also support many forest specialties, including Chinese Barbet, Blyth's Kingfisher and Fairy Pitta. Simple lodging is available in the reserve at Daqiutian station (大秋田), but a permit is needed for access.

■ Birdwatching ■

HUNAN PROVINCE

Dongting Lake Wetlands (洞庭湖湿地) Dongting Lake's wetlands support large numbers of wintering waterfowl, including much of the world's Lesser White-fronted Geese. Dongting is listed as a Ramsar Site and support a similar avifauna to Poyang, but much less visited. Other key species include Taiga and Tundra Bean Geese, Oriental Stork, Siberian Crane and Swan Goose. Access is from Hunan's provincial capital of Changsha or Yueyang city. A major birdwatching competition is held annually at the **East Dongting Nature Reserve** (东洞庭自然保护区) segment of the wetlands by Yueyang city.

Badagongshan (National) Nature Reserve (八大公山自然保护区) Despite its proximity to the world-famous **Wulingyuan World Heritage Site and Scenic Area** (武陵源风景区), Badagongshan is hardly visited, although it preserves large tracks of broadleaved evergreen and mixed forests, and supports a number of species more typical of the Sichuan uplands, including Spotted Laughingthrush and Temminck's Tragopan. Reeves's Pheasant has also been recorded recently. Much of the lower elevations are disturbed, and good forest occurs only above 1,800m. Access is from Sangzhi town.

FUJIAN PROVINCE

Minjiang Estuary (National) Wetland Park (闽江河口湿地自然保护区) Located at the Min River mouth, this park is one of the easiest places to see Chinese Crested Terns (spring–summer) and Spoon-billed Sandpipers (autumn–winter). There are large areas of reed beds, scrub and coastal flats that draw flocks of waders, ducks and gulls in winter and terns in summer. A permit is needed for access. Lodging is possible in Changle city.

Fuzhou (National) Forest Park and Botanical Gardens (福州国家森林公园) This well-forested park in the northern hills of Fuzhou, Fujian's provincial capital, is very popular among visiting birdwatchers. A number of sought-after species like Silver Pheasant, White-necklaced Partridge, Pale-headed Woodpecker and Spotted Wren Babbler (Elachura) are regulars in the park. Lodging is available in the park.

Emeifeng (Provincial) Nature Reserve (峨嵋峰自然保护区) This reserve in Taining county protects a section of the Wuyi Mountains along the Fujian–Jiangxi border, reaching over 1,700m at the densely forested Emeifeng. It is an excellent site for pheasants, including White-necklaced Partridge and Cabot's Tragopan, and Elliot's and Koklass Pheasants, and is now frequently visited by birdwatching tour groups. A permit is needed for access.

GUANGDONG PROVINCE

Sun Yat-sen University Guangzhou Campus (中山大学) Situated in Guangzhou, Guangdong's capital, the heavily wooded campus of Sun Yat-sen University is probably one of the best places to be introduced to the avifauna of south China, and supports a good selection of migrant passerines in autumn and early spring. Access is on Line 8 of the Guangzhou Metro. Alternatively, the **South China Botanical Gardens** (华南植物园) is another excellent site worth exploring.

Nanling (National) Forest Park (南岭国家森林公园) The most important site in

Guangdong for forest birds, this reserve protects part of the Nanling Mountains along the rugged Hunan–Guangdong border. The evergreen and bamboo forests support the highly localized Silver Oriole, and this is the best place to see the species in China. Other key species include Cabot's Tragopan, White-necklaced Partridge, Red-tailed Laughingthrush and Brown-chested Jungle Flycatcher. Access is from Ruyuan town.

Chebaling (National) Nature Reserve (车八岭自然保护区)
Originally established for conserving the South Chinese Tiger, Chebaling protects small, but good tracts of mostly hilly broadleaved evergreen forest near the Guangdong-Jiangxi border. Highlights include Blyth's Kingfisher, White-eared Night Heron and summering Fairy Pitta, as well as an avifauna representative of the South Chinese region. Access is from Shaoguan City.

Futian (National) Nature Reserve, Shenzhen (福田自然保护区, 深圳) Located by the Pearl River delta and opposite Hong Kong's **Mai Po Nature Reserve**, Futian protects parts of the Deep Bay wetland ecosystem, which supports more than 50,000 waterbirds in winter and spring. Black-faced Spoonbills, wintering ducks and shorebirds can be seen feeding on the mudflats at low tide. Access to the reserve requires a permit, but the adjacent **Mangrove Eco-Park** (红树林生态公园) is publicly accessible. Access is on Luobao Line on the Shenzhen Metro.

HAINAN (ISLAND) PROVINCE

Jianfengling (National) Forest Park (尖峰岭国家森林公园) All of Hainan's endemics and an excellent selection of lowland and montane forest species can be seen here, including many migrant passerines. The Mingfeng boardwalk (鸣风谷栈道) provides access to undisturbed forest conveniently behind the resort. Key species include Hainan Peacock Pheasant, Hainan Partridge and Rufous-cheeked Laughingthrush. Access is from Sanya city. **Bawangling (National) Nature Reserve** (霸王岭自然保护区) in Changjiang County, west Hainan, supports a very similar avifauna.

OPPORTUNITIES FOR NATURALISTS AND PHOTOGRAPHERS

Given the tendency for birdwatchers to visit a few popular sites, many of the remote parts of Southeast China remain poorly known. There are thus ample opportunities for new discoveries, especially site records that can improve knowledge of bird distributions and how these change with time. Carefully kept records of individual species, and daily lists compiled from birdwatching trips, are valuable and should be submitted to relevant databases (see page 171). Furthermore, birdwatchers can contribute to our understanding of bird ecology and behaviour if they keep detailed observations. Where possible, these observations should contain details on the number of individuals for each species recorded, age if apparent, sex (for dimorphic species), food items, interactions, elevation, geographical coordinates and surrounding habitat.

Besides a good pair of binoculars (8–10 x 42), birdwatchers should consider investing in a camera and sound-recording equipment. For portability, a digital camera with high optical magnification is useful, and models allowing up to 50x zoom are now widely available. Alternatively, digital single-lens reflex (DSLR) cameras can allow for higher image quality and versatility. Such set-ups should include at least a 300mm telephoto lens for adequate reach. For recording bird sounds, shotgun microphones are popular, but they are highly sensitive and should be muffled adequately with windscreens to reduce background noise. These microphones can be paired with a digital recorder using suitable powering cables.

Nomenclature and Taxonomy

Due to novel phylogenetic approaches, new insights in bird taxonomy are constantly emerging. This has led to frequent 'splits' where subspecies are elevated to the species level, or the converse in 'lumps', as well as periodic revisions of nomenclature. Such developments have not only increased the number of recognized species, but also challenged long-established, higher level relationships. For instance, Spotted Wren Babbler was only recently recognized as a highly distinct taxa in its own family, Elachuridae. For reasons of convenience, nomenclature, taxonomy and sequence of species in our book follow the latest version of the *Checklist of Birds of China* published by the China Bird Report (CBR) editorial group (see useful websites), and adopt the format and recommendations of the IOC World Bird List.

Common names of birds are less stable than scientific names, with large variations between countries in usage preferences. Again, we have chosen to adopt the English names recommended by the IOC checklist and Chinese names by the CBR checklist. In rare exceptions, for example in the case of newly recognized species, we have provided bird names most appropriate for the circumstance.

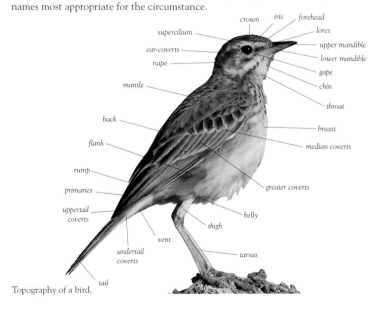

Topography of a bird.

Chinese Francolin
■ *Francolinus pintadeanus*
中华鹧鸪 (Zhōng huá zhè gū) 32cm

DESCRIPTION ssp. *pintadeanus*. Medium-sized partridge. Male (shown) strikingly patterned on head with rufous lateral crown-stripe, black brow and moustachial stripe, and white cheek-patch. Mantle rufous. Extensive white spotting on upperparts and underparts. Female mostly dull brown and heavily barred on underparts. **DISTRIBUTION** NE India, Southeast Asia to S and SE China. In SE China, from Guangdong to Zhejiang. **HABITS AND HABITATS** Uncommon to locally common resident in woodland, tree plantations, secondary scrub and grassland to 1,600m. Calls a series of five harsh, raspy notes, often from a fairly open position. **TIMING** All year round. **SITES** Suitable habitats across SE China (e.g. Datian, Hainan). **CONSERVATION** Least Concern.

Japanese Quail ■ *Coturnix japonica* 鹌鹑 (Ān chún) 18cm

DESCRIPTION Small, rounded and heavily streaked game bird. Male's face brownish, contrasting with long pale brow; rufous-washed on breast. Crown and most of upperparts brown with buff streaks and dark bars. Female (shown) paler faced with buff brow, bold

bands on neck sides and dark spotting on buff-washed breast. Similar female **King Quail** smaller and barred; not spotted on underparts. **DISTRIBUTION** Breeds Transbaikalia east to Russian Far East, NE China and Japan. Winters Korea, SW, E, SE and S China, mainland Southeast Asia. **HABITS AND HABITATS** Uncommon winter visitor and passage migrant, occurring in dry grassland, scrub and farmland, especially vegetation fringing ploughed areas. Shy; usually seen when flushed. **TIMING** Mostly Oct–Apr. **SITES** Poyang Lake (Jiangxi), Nanhui Dongtan (Shanghai), Hangzhou parks (Zhejiang). **CONSERVATION** Near Threatened. Has declined sharply in recent years due to habitat loss.

White-necklaced Partridge ■ *Arborophila gingica* 白眉山鹧鸪
(Bái méi shān zhè gū) 29cm

DESCRIPTION Distinctive, medium-sized partridge with thick black-and-chestnut band across lower neck, separated by thin white band. Crown chestnut, contrasting with white brow. Throat washed buff-orange with fine black streaking on neck sides. Upperparts brownish-grey; underparts mostly washed grey. Black spotting on rufous wings.
DISTRIBUTION SE China from Guangdong north to Zhejiang. Endemic to China. **HABITS AND HABITATS** Uncommon to locally common resident of broadleaved evergreen and mixed forests from 500 to 1,900m. Territorial duets consisting of a series of double whistles can often be heard from afar. **TIMING** All year round. **SITES** Emeifeng, Fuzhou Forest Park (Fujian), Wuyishan (Jiangxi), Wuyanling (Zhejiang). **CONSERVATION** Near Threatened. Affected mainly by habitat loss and fragmentation, and to some extent hunting.

Hainan Partridge ■ *Arborophila ardens*
海南山鹧鸪 (Hǎi nán shān zhè gū) 26cm

DESCRIPTION Small-sized partridge with distinct facial patterns. Black facial mask with prominent white spot on ear-coverts; contrasts with orange-red lower throat. Pale brow extends to nape. Upperparts brown with black barring; underparts washed orange-buff with white streaking on flanks. **DISTRIBUTION** Endemic to Hainan Island. **HABITS AND HABITATS** Rare to locally common resident of tropical evergreen forests from 400 to 1,300m. Forages quietly on forest floor in small family groups. Pairs often heard giving a territorial duet that comprises a series of double whistles, especially in early mornings and evenings. **TIMING** All year round. **SITES** Bawangling, Diaoluoshan, Jianfengling, Yinggeling (Hainan). **CONSERVATION** Vulnerable. Class I protected species. Threatened mainly by habitat loss and degradation, and hunting.

Chinese Bamboo Partridge ■ *Bambusicola thoracicus* 灰胸竹鸡
(Huī xiōng zhú jī) 33cm

DESCRIPTION ssp. *thoracicus*. Medium-sized, colourful partridge with a longish tail. Bluish-grey forehead and brow extending to nape, against rufous-brown face and neck.

Upperparts greyish-brown with chestnut spots; underparts washed buff-yellow with bold black spotting on breast and flanks. **DISTRIBUTION** C, E and SE China, where widespread; also Taiwan. Endemic to China, introduced to Japan. **HABITS AND HABITATS** Common resident of disturbed forests, forest edges, bamboo and woodland, from lowlands to about 1,100m. Seen in pairs or small family groups. Active, territorial song often described as a loud *people pray, people pray, people pray*, becoming slower towards the end. **TIMING** All year round. **SITES** Fuzhou Forest Park (Fujian), Nanling (Guangdong), Tianmushan (Zhejiang). **CONSERVATION** Least Concern.

Cabot's Tragopan ■ *Tragopan caboti* 黄腹角雉
(Huáng fù jiǎo zhì) M 61cm, F 50cm

DESCRIPTION ssp. *caboti*. Medium-sized pheasant with a short tail. Male distinctive: upperparts reddish-brown with bold, buff spotting; underparts washed yellowish-buff.

Head black with orange facial skin. Red and bluish facial lappets and orange fleshy 'horns' are erected during courtship display'. Female mostly brownish-grey with extensive streaking on underparts. **DISTRIBUTION** C, S and SE China from Guangdong to Zhejiang. Endemic to China. **HABITS AND HABITATS** Rare to locally common resident of broadleaved and mixed evergreen forests from 600 to 1,800m. Besides well-known courtship display, sometimes engages in fights with much kicking. Nest constructed

with branches and mosses up in trees. Calls include a harsh series of clucks. **TIMING** All year round. **SITES** Emeifeng, Longqishan (Fujian), Nanling (Guangdong), Wuyishan (Jiangxi), Wuyanling (Zhejiang). **CONSERVATION** Vulnerable. Class I protected species. Threatened mainly due to habitat loss.

Koklass Pheasant ■ *Pucrasia macrolopha* 勺鸡 (Sháo jī) M 61cm, F 54cm

DESCRIPTION ssp. *darwini*. Medium-sized pheasant with a short tail. Male's (shown) head dark green with a white patch on neck sides; long ear-tufts. Narrow chestnut patch from breast to belly. Plumage mostly grey-buff, appearing finely streaked due to black shaft streaks of plumes. Female smaller and duller than male, also with white patch on neck sides.

DISTRIBUTION NW Himalayas east to across much of C, SE and E China. In SE China, from Fujian to Zhejiang. **HABITS AND HABITATS** Locally common to rare resident of mixed and coniferous forests from 600 to 1,500m, keeping to dense, shrubby vegetation. Usually seen alone or in pairs. Calls a harsh series of *kak-krrrk*, especially in early morning. **TIMING** All year round. **SITES** Emeifeng (Fujian), Wuyishan (Jiangxi), Qingliangfeng (Zhejiang). **CONSERVATION** Least Concern. Class II protected species.

Red Junglefowl ■ *Gallus gallus* 红原鸡 (Hóng yuán jī) M 70cm, F 42cm

DESCRIPTION ssp. *jabouillei*. Wild ancestor of domestic chicken, but larger and more colourful. Male has a red comb, throat lappet and facial skin. Hackles and uppertail-coverts rich golden-orange. Underparts, wing-coverts and tail feathers iridescent bluish-green. Female (shown) mostly dull brown with pink facial skin. **DISTRIBUTION** N, E and NE Indian subcontinent to S China and Southeast Asia. In SE China, from Hainan to S Guangdong. **HABITS AND HABITATS**

Uncommon resident of disturbed areas in broadleaved evergreen forests, forest edges and scrub, from lowlands to 1,000m. Males usually seen alone or with groups of females. Forages on ground, but flies up to roost on trees or when disturbed. **TIMING** All year round. **SITES** Tongledashan (Guangdong), Datian, Jianfengling (Hainan).

CONSERVATION Least Concern. Class II protected species. Threatened by hybridization with domestic fowl.

Silver Pheasant ■ *Lophura nycthemera* 白鹇 (Bái xián) M 110cm, F 94cm

DESCRIPTION ssp. *fokiensis* (shown), *nycthemera* and *whiteheadi* (Hainan). Large white pheasant with black underparts. Male's upperparts, wings and tail mostly white with fine

vermiculations, contrasting with black underparts from throat to vent, and black crown and crest. Female shaped similarly to male, but mostly brown and tail shorter. Both sexes have red facial skin. **DISTRIBUTION** Mainland Southeast Asia to S and SE China. Widespread SE China. **HABITS AND HABITATS** Fairly common resident of broadleaved evergreen and mixed forests from 100 to 2,150m. Usually

in small groups with one dominant male and several females. When disturbed, quietly walks away or flushes to trees. **TIMING** All year round. **SITES** Fuzhou Forest Park (Fujian), Jianfengling (Hainan), Wuyishan (Jiangxi), Gutianshan (Zhejiang). **CONSERVATION** Least Concern. Class II protected species.

Elliot's Pheasant ■ *Syrmaticus ellioti* 白颈长尾雉 (Bái jǐng cháng wěi zhì) M 80cm, F 50cm

DESCRIPTION Large-sized, colourful pheasant with a long tail. Male's (shown) head and hood greyish-white; back, mantle to breast deep chestnut with black spotting. Wings

rich chestnut with two white bars. Tail long with alternating chestnut and white bars. Female duller than male; shorter tailed; mostly greyish-brown with prominent greyish-white neck sides. **DISTRIBUTION** C and SE China from Hunan to Zhejiang. Endemic to China. **HABITS AND HABITATS** Locally common to rare resident of broadleaved evergreen and mixed forests from 200 to 1,500m, foraging in thick shrub or bamboo cover. Occurs singly or in small groups. **TIMING** All year round. **SITES** Emeifeng, Longqishan (Fujian), Guanshan, Wuyuan (Jiangxi), Gutianshan (Zhejiang). **CONSERVATION** Near Threatened. Class I protected species. Threatened mainly by habitat loss.

Hainan Peacock Pheasant ■ *Polyplectron katsumatae* 海南孔雀雉
(Hǎi nán kǒng què zhì) M 51cm, F 38cm

DESCRIPTION Small-sized, brownish-grey pheasant. Throat of male (shown) white, facial skin and bill brick-red; upperparts and tail patterned with fine white spotting and large green ocelli. Female resembles male,

but is smaller and less strongly marked.
DISTRIBUTION Endemic to Hainan Island.
HABITS AND HABITATS Uncommon to rare resident of broadleaved evergreen forests from 600 to 1,300m. Usually seen singly or in pairs foraging quietly on forest floor for arthropods, keeping to compact territories of about 3 ha. Calls various harsh *krrrk* and *kok* notes. **TIMING** All year round. **SITES** Bawangling, Jianfengling, Yinggeling (Hainan). **CONSERVATION** Endangered. Class I protected species. Global population estimated at 250–1,000 mature individuals. Threatened mainly by habitat loss and illegal trapping.

Swan Goose ■ *Anser cygnoides* 鸿雁 (Hóng yàn) 87cm

DESCRIPTION Large but slender goose with a longish bill and neck. Crown and entire hind neck dark brown, contrasting strongly with pale neck and breast. Bill black with white base on adults. Legs bright orange.

DISTRIBUTION Breeds Mongolia, Transbaikalia to NE China, and S Russian Far East. Winters south to E, SE China and Korea, rarely Japan and Taiwan.
HABITS AND HABITATS Occurs on migration and during winter in wetlands, especially those adjacent to grassland or with flooded vegetation. Often forms large mixed flocks with other goose species.
TIMING Oct–Feb. **SITES** Poyang Lake, Nanjishan (Jiangxi), Minjiang Estuary (Fujian). **CONSERVATION** Vulnerable. Population in decline due to habitat loss and illegal trapping. Virtually the entire global population of over 60,000 winters in the lower Yangtze floodplain, especially Poyang Lake.

Taiga Bean Goose ■ *Anser fabalis* 豆雁 (Dòu yàn) 90cm

DESCRIPTION Split from **Tundra Bean Goose**. Head and neck dark brown, contrasting with buff-washed lower neck and breast. Upperparts dark brown. Bill black with orangey-

yellow patch before black tip. Longer neck, more slender bill and larger in size than **Tundra Bean Goose**. **DISTRIBUTION** Breeds NE Asia from Transbaikalia and Siberia south to Amur River. Winters SE China especially along Yangtze valley, Korea and Japan; rarely Taiwan. **HABITS AND HABITATS** Locally common winter visitor occuring on lakes, freshwater marshes and open farmland with flooded vegetation, stages on river during migration. Forms large flocks with other goose species. **TIMING** Oct–Feb. **SITES** Dongting Lake (Hunan), Poyang Lake (Jiangxi). **CONSERVATION** Least Concern. Global population appears to be declining, but situation in China unclear.

Tundra Bean Goose ■ *Anser serrirostris* 短嘴豆雁
(Duǎn zuǐ dòu yàn) 85cm

DESCRIPTION Head and neck dark brown, contrasting with buff-washed lower neck and breast. Upperparts dark brown. As in **Taiga Bean Goose**, bill black with orange

patch near tip and black tip, but stumpier, giving a 'sad' look. Smaller in size than **Taiga Bean Goose**; neck thicker and shorter, and head appears more rounded. **DISTRIBUTION** Breeds on Arctic tundra from NE Siberia, east to Chukotka and W Kamchatka. Winters in E China, Korea and Japan. **HABITS AND HABITATS** Common winter visitor, occuring on lakes, flooded grassland and freshwater marshes; also grazes on open farmland. Forms large flocks with other goose species. **TIMING** Oct–Feb. **SITES** Dongting Lake (Hunan), Poyang Lake (Jiangxi). **CONSERVATION** Least Concern, but populations appear to be declining.

Greater White-fronted Goose ▪ *Anser albifrons* 白额雁
(Bái é yàn) 72cm

DESCRIPTION Medium-sized dark goose with a short neck and compact body. Bright pink bill against white face and forehead-patch. Upperparts greyish-brown; underparts
mostly pale grey with black speckling on belly. Juveniles darker on face than adults, and lack bars on belly and white face. Similar to **Lesser White-fronted Goose**, but larger and lacks yellow eye-ring. **DISTRIBUTION** Breeds across Arctic tundra of Russia, Alaska and NC North America to Greenland. Winters south to S North America, Europe, across to E Asia. **HABITS AND HABITATS** Common winter visitor, occurring on freshwater wetlands, especially lakes with adjacent wet grassland and open farmland. Forms large flocks in excess of 10,000 individuals at favoured wintering sites, especially Poyang Lake. **TIMING** Oct–Feb. **SITES** Poyang Lake (Jiangxi), Dongting Lake (Hunan). **CONSERVATION** Least Concern. Class II protected species.

Lesser White-fronted Goose ▪ *Anser erythropus* 小白额雁
(Xiǎo bái é yàn) 58cm

DESCRIPTION Similar to **Greater White-fronted Goose** but smaller, neck and bill shorter and head more rounded. White patch at base of pink bill wider and extends to crown; yellow eye-ring. Black speckling on belly less distinct than in **Greater White-fronted Goose**. Juveniles lack black belly speckling and white facial patch. **DISTRIBUTION** Breeds Arctic tundra of N Europe and N Russia, east to Chukotka. Winters south to Europe and Middle East, eastwards to Japan, Korea and E, SE China mostly in Yangtze basin. **HABITS AND HABITATS** Locally common to rare winter visitor to freshwater wetlands, lakes with adjacent grassland and open farmland. On migration often in small groups or singles within flocks of other geese like **Greater White-fronted Goose**. **TIMING** Oct–Feb. **SITES** Dongting Lake (Hunan), Poyang Lake (Jiangxi). **CONSERVATION** Vulnerable. Threatened due to hunting and habitat loss. Approximately 10,000 wintering pairs in the East Asian-Australasian Flyway, mostly at Dongting Lake.

Tundra Swan ▪ *Cygnus columbianus* 小天鹅 (Xiǎo tiān é) 120–150cm

DESCRIPTION ssp. *bewickii*. Adult mostly white; some show rusty wash on face, neck and breast. Bill mostly black with rounded, somewhat variable yellow base. Similar to **Whooper**

Swan, but smaller and stumpier, and with a shorter neck and more rounded head. Juveniles dull greyish-brown with pink-based black bill. **DISTRIBUTION** Breeds Arctic tundra across North America and N Russia. Eurasian populations winter patchily across Europe, Middle East to E Asia, where mostly Japan, Korea, NE and E China. **HABITS AND HABITATS** Locally common to uncommon winter visitor to lakes, rivers and freshwater wetlands; also coastal mudflats. Visits to rest and feed in cultivated areas during migration. Flocks in flight form linear or 'V' patterns. **TIMING** Oct–Feb. **SITES** Minjiang Estuary (Fujian), Dongting Lake (Hunan), Poyang Lake (Jiangxi). **CONSERVATION** Least Concern. Class II protected species.

Mandarin Duck ▪ *Aix galericulata* 鸳鸯 (Yuān yāng) 46cm

DESCRIPTION Colourful duck with a large head and longish tail. Males vividly coloured, with a broad white stripe from bill base to end of crest, contrasting with orange head

sides and 'whiskers'. Upper breast iridescent blue, bounded on flanks by black-and-white bars; flanks orange-brown. Inner webs of orange tertials rise like 'sails'. Female mostly greyish-brown with bold white spots on underparts; also white eye-ring, bill base and throat. **DISTRIBUTION** Breeds Russian Far East, NE China, Korea, N Japan and Taiwan. Winters south to Japan, E, SE China. Has occasionally bred SE China. Introduced to W Europe. **HABITS AND HABITATS** Locally common winter visitor, occurring on lakes and rivers with well-vegetated banks, sometimes in very large groups. Nests in tree-holes; on migration congregates to form flocks. **TIMING** Mostly Oct–Mar. **SITES** Wuyuan (Jiangxi), Hangzhou West Lake (Zhejiang). **CONSERVATION** Least Concern. Class II protected species.

Baer's Pochard ▪ *Aythya baeri* 青头潜鸭 (Qīng tóu qián yā) 45cm

DESCRIPTION Small dark duck with a rounded head. Male with glossy green head and neck; breast reddish-brown, flanks brownish. Male has yellow eyes, female (shown) brown. Both sexes when non-breeding appear dark headed and less strongly marked. Female similar to rarer **Ferruginous Duck**, but head rounder and flanks whitish.

DISTRIBUTION Breeds NE China and S Russian Far East. Winters across E China south of Yangtze River, to NE India and Southeast Asia; also Taiwan (rare). **HABITS AND HABITATS** Uncommon to rare winter visitor to large lakes with vegetated margins and freshwater marshes. Forms mixed flocks with other diving ducks (e.g. Tufted Duck). **TIMING** Nov–Mar. **SITES** Dongting Lake (Hunan), Poyang Lake (Jiangxi), Xixi National Wetland Park (Zhejiang). **CONSERVATION** Critically Endangered. Population declining rapidly and estimated to be as low as 150 mature individuals.

Scaly-sided Merganser ▪ *Mergus squamatus* 中华秋沙鸭
(Zhōng huá qiū shā yā) 57cm

DESCRIPTION Green-headed sawbill with well-marked underparts. Male (shown): head to neck sides dark glossy green with long double crest. Scapulars and flight feathers black, contrasting with white coverts. Flank feathers dark-edged, appearing heavily scaled, and unlike white underparts of **Common Merganser**. Females orange-brown on head, more extensive and finely vermiculated scaling, from flanks to neck sides. **DISTRIBUTION** Breeds Russian Far East and NE China. Winters C, E, SE China, including Taiwan; also Korea and Japan. **HABITS AND HABITATS** Locally uncommon winter visitor to fast-flowing rivers and lakes with forested banks, from lowlands to 1,000m. Usually singly or small mixed-sex groups. Dives regularly to catch large insects or small fish. Often associates with Mandarin Ducks. **TIMING** Nov–Mar. **SITES** Wuyuan (Jiangxi) **CONSERVATION** Endangered. Class I protected species.

Great Crested Grebe ▪ *Podiceps cristatus* 凤头䴙䴘 (Fèng tóu pì tī) 49cm

DESCRIPTION ssp. *cristatus*. Large grebe with a long, thin neck. Pink bill long and pointed. Breeding plumage (shown) unmistakable, with black crest and long, orange-

brown ear-tufts extending to nape. Non-breeding birds paler, lack ear-tufts, face white and upperparts greyish-brown. **DISTRIBUTION** Breeds W Europe to Tibetan Plateau and Russian Far East; south to C and E China; also Australia. Winters south to Africa, Indian subcontinent and East Asia, including SE China. **HABITS AND HABITATS** Fairly common winter visitor, occurring on large water bodies like lakes, rivers and reservoirs, and in estuarine wetlands and coastal areas. Dives regularly for fish and crustaceans, often surfacing further away from initial point of dive. **TIMING** Oct–Mar. **SITES** Futian (Guangdong), Nanhui Dongtan (Shanghai). **CONSERVATION** Least Concern.

Oriental Stork ▪ *Ciconia boyciana* 东方白鹳 (Dōng fāng bái guàn) 110cm

DESCRIPTION Large, distinctive white stork with a black bill. Plumage mostly white with contrasting black flight feathers and wing-coverts. Facial skin and legs red. **DISTRIBUTION**

Breeds NE China and Russian Far East, recently E (Liaoning), SE China (Jiangxi). Winters S and SE China, mostly in lower Yangtze River basin; also Japan and Korea. **HABITS AND HABITATS** Locally common to uncommon winter visitor occurring on large water bodies like lakes with well-vegetated margins and freshwater wetlands; small numbers occur in coastal wetlands. In recent years, found breeding on electric pylons in Poyang Lake area, indicating significant southwards extension of breeding range. **TIMING** Mostly Nov–Mar. **SITES** Dongting Lake (Hunan), Poyang Lake (Jiangxi), Nanhui Dongtan (Shanghai). **CONSERVATION** Endangered. Population estimated at 3,000 individuals. Threatened by habitat loss and degradation, poisoning and illegal trapping.

Black-faced Spoonbill ▪ *Platalea minor* 黑脸琵鹭 (Hēi liǎn pí lù) 70cm

DESCRIPTION Medium-sized, all-white waterbird with spatulate-shaped bill. Similar to larger **Eurasian Spoonbill**, but adult (shown) has black facial skin and bill, and lacks yellowish throat. Juveniles have greyer facial skin and paler bill than adults; also show black wing-tips in flight. **DISTRIBUTION** Breeds on islands off Yellow Sea coast in Korea and NE China; also Russian Far East. Winters N Vietnam, S Japan, S and SE China coast and Taiwan. **HABITS AND HABITATS** Locally common to uncommon winter visitor to coastal and estuarine mudflats; occasionally found further inland. Flight a distinct flap-glide pattern, unlike that of egrets. **TIMING** Oct–Mar. **SITES** Chongming Dongtan (Shanghai), Fuqing (Fujian), Futian and Haifeng coast (Guangdong). **CONSERVATION** Endangered. Class II protected species. Fewer than 300 individuals in late 1980, but population has recovered to 2,700 individuals. Threats include (wintering) habitat loss and human disturbance. Wetlands along Deep Bay in Hong Kong and Shenzhen support one of largest concentrations of world population outside Taiwan.

Chinese Egret ▪ *Egretta eulophotes* 黄嘴白鹭 (Huáng zuǐ bái lù) 68cm

DESCRIPTION Medium-sized white egret similar to **Little** and **Pacific Reef Egrets**. Breeding birds orange billed with blue facial skin and shaggy crest; legs dirty green with yellow feet. Non-breeding birds have two-toned, dagger-shaped bills with pink to yellowish lower mandible; legs and feet dirty green. **DISTRIBUTION** Breeds on islets off Russian Far East, Korea, E and S China coasts, south to Fujian (e.g. Caiyu, Kinmen). Winters Southeast Asia (e.g. Philippines). **HABITS AND HABITATS** Uncommon summer breeder and passage migrant, occurring on coastal mudflats, sandy beaches and occasionally reefs. Breeds in small colonies on islands of Fujian, Zhejiang coasts. Active feeder, often walking swiftly over soft substrates before squatting and tilting body forwards to stab prey. Foraging strategies very different from those of other egrets. **TIMING** Apr–Sep. **SITES** Minjiang Estuary (Fujian), Futian (Guangdong), Nanhui Dongtan (Shanghai). **CONSERVATION** Vulnerable. Class II protected species. Breeding population estimated at no more than 10,000 individuals.

Great Egret ■ *Ardea alba* 大白鷺
(Dà bái lù) 95cm

DESCRIPTION Largest of white egrets. Longest neck among egrets; neck distinctly kinked. Non-breeding birds (shown) yellow on bill and facial skin; legs black. Breeding birds black billed and show bluish facial skin. **DISTRIBUTION** Breeds widely across Americas, Eurasia, Africa and Australia. Many northern populations migrate to winter in tropics. **HABITS AND HABITATS** Common summer breeder and winter visitor, occurring on coastal mudflats, mangroves, paddy fields, lakes with well-vegetated fringes and wet grassland. Seen singly or in small numbers, but mixes with other egrets when foraging. **TIMING** All year round. **SITES** Suitable habitat across SE China. **CONSERVATION** Least Concern.

White-eared Night Heron ■ *Gorsachius magnificus* 海南鳽
(Hǎi nán jiān) 58cm

DESCRIPTION Medium-sized heron with large eyes. Head and crest black, with white patch behind eye and thin white line below yellow facial skin. Thick black stripe along neck sides, contrasting with rufous neck-patch. Upperparts and wings dark brown; underparts heavily flecked buff. Juveniles resembles adult, but extensively spotted on upperparts. **DISTRIBUTION** N Vietnam; SE China, from Hainan to Zhejiang. **HABITS AND HABITATS** Rare to uncommon resident of broadleaved evergreen forests and occasionally tree plantations, usually near rivers or lakes; also known to forage in paddy fields near forests. Feeds on fish and crabs at night, dispersing to feeding grounds at dusk. Sedentary and does not form breeding colonies. **TIMING** All year round. **SITES** Chebaling (Guangdong), Jiulianshan (Jiangxi), Qiandao Lake (Zhejiang). **CONSERVATION** Endangered. Class II protected species. Threatened by habitat loss and illegal trapping.

juvenile

Chinese Pond Heron ▪ *Ardeola bacchus* 池鷺 (Chí lù) 49cm

DESCRIPTION Stocky, short-necked heron. Breeding birds have a rich chestnut head and neck, and a greyish-black mantle. Non-breeding birds streaked brown on pale buffy head, neck and breast; wings white and mantle brown. Bill yellowish with a black tip. In flight, white wings and tail show prominently. **DISTRIBUTION** Breeds N Southeast Asia and S China, to Russian Far East, NE China and Japan. Northern populations winter S China and Southeast Asia.

breeding

HABITS AND HABITATS Very common summer breeder, passage migrant and winter visitor. Occurs in coastal mudflats, estuaries, freshwater marshes and paddy fields. Solitary or in loose flocks. Calls a harsh, nasal *gaa*. **TIMING** All year round. **SITES** Suitable habitat across SE China. **CONSERVATION** Least Concern.

Yellow Bittern ▪ *Ixobrychus sinensis* 黄苇鳽 (Huáng wěi jiān) 37cm

DESCRIPTION Small, pale buffy-brown bittern. Female (shown) lacks black crown of males. In flight, black flight feathers contrast with buff wing-coverts. Juvenile heavily streaked on back and underparts. Similar-looking adult **Von Schrenk's Bittern** is rich chestnut on upperparts, not buff. **DISTRIBUTION** Breeds N Indian subcontinent, Southeast Asia and much of East Asia to Russian Far East. E Asian populations migrate to winter in Southeast Asia. **HABITS AND HABITAT** Common summer breeder and winter visitor. Occurs in freshwater marshes, paddy fields, wet grassland and inland wetlands, including reed-fringed canals and well-vegetated lakes. Solitary. Often freezes with neck outstretched when it senses danger. Utters a series of harsh *kaark* notes. **TIMING** All year round. **SITES** Suitable habitat across SE China. **CONSERVATION** Least Concern.

Cinnamon Bittern ▪ *Ixobrychus cinnamomeus* 栗苇鳽 (Lì wěi jiān) 39cm

DESCRIPTION Small, richly coloured bittern. Male (shown) rich cinnamon-brown on upperparts; underparts pale buff. Female darker brown on upperparts with fine speckling, and heavily streaked on underparts. In flight, lacks black flight feathers of other *Ixobrychus* bitterns. **DISTRIBUTION** Breeds Indian subcontinent, Southeast Asia and much of coastal East Asia. E Asian populations migrate to winter in Southeast Asia. **HABITS AND HABITATS** Common resident and winter visitor. Occurs in freshwater marshes, wet grassland, paddy fields and sometimes woodland. Solitary and secretive; hides in dense vegetation and usually seen in flight when flushed. Calls a quick series of harsh *krk* notes. **TIMING** All year round. **SITES** Suitable habitat across SE China. **CONSERVATION** Least Concern.

Von Schrenk's Bittern ▪ *Ixobrychus eurhythmus* 紫背苇鳽 (Zǐ bèi wěi jiān) 36cm

DESCRIPTION Small, richly coloured bittern. Adult male (shown) rich chestnut on upperparts, contrasting with buff-brown underparts and wing-coverts. Female and young spotted white on upperparts; underparts streaked brown. Similar **Cinnamon Bittern** mostly orangey-brown, and lacks buff wing-coverts. **DISTRIBUTION** Breeds SE, E China

to Russian Far East and Japan. Winters S, SE China and much of Southeast Asia. **HABITS AND HABITATS** Uncommon passage migrant, very rare summer breeder. Occurs in freshwater marshes, paddy fields and forested streams (wintering birds). Shy, often hiding in dense vegetation. Calls a grating *krek*. Where it breeds, calls a repetitive, low *ooo*. **TIMING** Mostly May–Jun, Sep–Oct. **SITES** Suitable habitat across SE China. **CONSERVATION** Least Concern, but in decline – some populations in East Asia have disappeared.

Dalmatian Pelican ■ *Pelecanus crispus* 卷羽鹈鹕 (Juán yǔ tí hú) 170cm

DESCRIPTION Very large waterbird with unmistakable silhouette – massive head with curly head plumes, long bill, large body and stumpy legs. Plumage mainly greyish-white, mostly covering black outer primaries; pouch pinkish-yellow. Juveniles mostly greyish with pale pouch. **Great White** and **Spot-billed Pelicans** have more extensive bare facial skin, and more black flight feathers. **DISTRIBUTION** Breeds E Europe eastwards to W Mongolia.

Winters discontinuously SE Europe, to W Indian subcontinent and SE China. **HABITS AND HABITATS** Uncommon to rare winter visitor, occurring on large water bodies (freshwater lakes) and coastal mudflats. Gregarious; usually seen in small parties. **TIMING** Dec–Mar. **SITES** Minjiang Estuary (Fujian), Haifeng (Guangdong), Poyang Lake (Jiangxi), Wenzhou Bay (Zhejiang). **CONSERVATION** Vulnerable. Class II protected species. East Asian population numbers around 100 individuals and is in decline. Loss of wetland habitats and hunting are key threats.

Great Cormorant ■ *Phalacrocorax carbo* 普通鸬鹚 (Pǔ tōng lú cí) 85cm

DESCRIPTION ssp. *sinensis*. Large black cormorant with a long neck. Underparts of breeding adults (shown) black; wings and scapulars glossy green; head sides to neck white. Facial skin white; throat yellow. Immatures have whitish belly and lack glossy feathers on wings. **DISTRIBUTION** Widespread but patchily distributed across Eurasia, North America and Australia. Some populations migratory. E Asian populations breed Mongolia, Russian Far East, N China, and winter S, SE China and Southeast Asia. **HABITS AND HABITATS** Locally common to uncommon winter visitor to lakes, rivers and reservoirs; also fishponds, estuarine areas and coastal wetlands. Forages for fish mainly in early morning; preens and engages in other social interactions during rest of day. Gregarious. **TIMING** Mostly Oct–Mar. A few oversummer. **SITES** Poyang Lake, Wuyuan (Jiangxi), Xiamen, Futian (Fujian), Nanhui Dongtan (Shanghai). **CONSERVATION** Least Concern. Sometimes persecuted by fish farmers at fish ponds.

Black Baza ▪ *Aviceda leuphotes* 黑冠鹃隼
(Hēi guān juān sǔn) 32cm

DESCRIPTION ssp. *syama*. Small pied raptor with erect crest. Upperparts mostly black, with prominent white breast-patch; chestnut barring on belly. In flight, wings broad, appearing 'rounded' at the ends, and narrower at the bases. **DISTRIBUTION** Breeds Himalayan foothills, S and NE India, parts of Southeast Asia and much of SE China. SE Chinese populations migrate to winter across Southeast Asia. **HABITS AND HABITATS** Uncommon summer breeder and passage migrant. Occurs in lowland and broadleaved evergreen forests, and woodland from lowlands to 1,850m. Predator of large insects and lizards, hunting in small groups in forest canopy. Calls a three-note whistle, the last note inflected upwards. **TIMING** Apr–Oct. **SITES** Chebaling, Dinghushan (Guangdong), Wuyishan, Wuyuan (Jiangxi). **CONSERVATION** Least Concern. Class II protected species. Populations in decline due to habitat loss in wintering areas.

Crested Honey Buzzard ▪ *Pernis ptilorhynchus* 凤头蜂鹰
(Fèng tóu fēng yīng) 54-65cm

DESCRIPTION ssp. *orientalis*. Medium-sized raptor with variable plumage. Head small with a slight crest, broad wings and long, often fanned out tail. Adults (shown) brown with a white throat-patch; finely streaked on breast; banded belly. Dark-morph birds dark brown with a white throat-patch. Underwing pattern similar to that of pale morph, but more strongly marked. **DISTRIBUTION** Breeds Indian subcontinent, NE and E Asia, east to Southeast Asia. Populations in Russian Far East, NE China and Korea migrate to winter in Southeast Asia. Widespread across SE China. **HABITS AND HABITATS**

Uncommon passage migrant and rare winter visitor; also rare summer breeder. Occurs in broadleaved evergreen and mixed forests, forest edges and woodland from lowlands to 1,800m, also occasionally urban parks on migration. Specialized feeder on bee larvae and beeswax obtained by tearing beehives. Call a thin, airy whistle. **TIMING** Mostly Apr–May, Sep–Oct. **SITES** Suitable habitats across SE China. Migration path across region poorly known. **CONSERVATION** Least Concern. Class II protected species.

Black Kite ▪ *Milvus migrans* 黑鸢 (Hēi yuān) 58-65cm

DESCRIPTION ssp. *lineatus (shown) formosanus (Hainan)*. Large dark raptor with a long, shallowly forked tail. Plumage brown; paler on head with a dark patch over ear-coverts. Underparts boldly streaked. In flight note pale patch on bases of primaries. Juveniles more heavily streaked than adults. **DISTRIBUTION** Breeds across Eurasia: N Europe east to Russia, E and SE China. Northern populations winter further south, including SE China. Also Africa and Indian subcontinent. **HABITS AND HABITATS** Common winter visitor and resident across various habitats, including farmland, wetlands, open country and even urban areas, especially near coasts. Usually in small groups, but sometimes seen singly. An adaptable scavenger, but occasionally catches live prey. Roosts in large flocks (up to several hundred birds). **TIMING** All year round; commoner in winter due to influx of migrants from north. **SITES** Suitable habitat in SE China. **CONSERVATION** Least Concern. Class II protected species. The most conspicuous raptor across the region.

Crested Serpent Eagle ▪ *Spilornis cheela* 蛇雕 (Shé diāo) 50–70cm

DESCRIPTION ssp. *ricketti, rutherfordi (Hainan)*. Large, distinctive hawk with a black crest and yellow face. Plumage brown with underparts spotted; tail white with broad black terminal band. In flight, underwing-coverts and flight feathers banded black and white. Juveniles pale, with spotting on nape and back. **DISTRIBUTION** Indian subcontinent, S China and Southeast Asia. Widespread in SE China from Hainan to N Zhejiang. **HABITS AND HABITATS** Uncommon resident of broadleaved evergreen forests, forest edges, tree plantations and occasionally farmland. Specialist predator of arboreal reptiles, particularly snakes. Calls 3–4 shrill, airy whistles. **TIMING** All year round. **SITES** Chebaling (Guangdong), Jianfengling (Hainan), Jiulianshan (Jiangxi). **CONSERVATION** Least Concern. Class II protected species. Widespread, but likely affected by deforestation.

hainanus

Eastern Marsh Harrier ▪ *Circus spilonotus* 白腹鹞 (Bái fù yào) 43-56cm

DESCRIPTION Medium-sized, long-tailed harrier. Male blackish-brown on mantle, back and wings, tail white; heavily streaked from neck to belly. In flight, mostly greyish-white

below with dark wing-tips and streaked underparts. Female brown with streaking from neck to upper belly; belly and thigh rufous. In flight, note centrally broken thin bars on upper tail, unlike in **Pied Harrier**. Juveniles (shown) mostly dark brown and pale headed. **DISTRIBUTION** Breeds E Russia, NE China. Winters across E, SE, S China, Taiwan, Korea and Japan. Also Southeast Asia. **HABITS AND HABITATS** Fairly common winter visitor, occurring in wet grassland, open scrub, freshwater marshes and farmland. Hunts by quartering low in search of ground prey (e.g. rodents and frogs). **TIMING** Sep–Mar. **SITES** Suitable habitat across SE China. **CONSERVATION** Least Concern. Class II protected species.

Pied Harrier ▪ *Circus melanoleucos* 鹊鹞 (Què yào) 44-50cm

DESCRIPTION Striking black-and-white harrier. Male black from head to breast, and on much of upperparts; also black on median upperwing-coverts and primaries; rest of underparts white; tail grey. In flight, underwings and white body contrast with black hood.

Female dark brown on upperparts, with pale patch on shoulders and mostly grey flight feathers. Young birds mostly rufous-brown on underparts and wing-coverts. In flight, note five unbroken bars on tail. **DISTRIBUTION** Breeds in E Mongolia, E Russia to NE China. Winters NE, E India, Southeast Asia and S, SE China. **HABITS AND HABITATS** Uncommon winter visitor and

passage migrant, occurring in wet and dry grasslands, freshwater wetlands, open scrub and farmland (e.g. paddy fields). Like other harriers, quarters low for ground prey. **TIMING** Nov–Apr. Autumn passage in October. **SITES** Yanghu Wetlands (Hunan), Nanhui Dongtan (Shanghai). **CONSERVATION** Least Concern. Class II protected species.

juvenile

Crested Goshawk ▪ *Accipiter trivirgatus* 凤头鹰 (Fèng tóu yīng) 40-46cm

DESCRIPTION ssp. *indicus*. Chunky, medium-sized raptor with a short crest. Head grey with a distinct throat-stripe; upperparts dark brown. Juveniles (shown) are brown-headed. Breast streaked and belly barred rufous. In flight, note three black bands across flight feathers and alternating broad black/pale bands on tail. **DISTRIBUTION** Himalayan foothills, E, S India, S, SE China and much of Southeast Asia eastwards to Philippines. In SE China, from Hainan to Zhejiang. **HABITS AND HABITATS** Uncommon resident of broadleaved evergreen and mixed forests, woodlands, tree plantations and occasionally parkland. Specialist predator of birds and small mammals. Calls a repeated, drongo-like *chweet*. **TIMING** All year round. **SITES** Bawangling (Hainan), Chebaling (Guangdong). **CONSERVATION** Least Concern. Class II protected species.

immature

Chinese Sparrowhawk

▪ *Accipiter soloensis* 赤腹鹰 (Chì fù yīng) 28-35cm

DESCRIPTION Small pale sparrowhawk. Adult (shown) grey on upperparts, white on underparts, with faint orange barring on breast; also bright orange cere. In flight, black tips to primaries diagnostic, distinguishing it from other sparrowhawks. Young birds dark brown on upperparts; streaked heavily from throat to breast, while lower belly is barred. In flight, note faint banding of flight feathers with dark tips of wings. **DISTRIBUTION** Breeds in E, NE, SE China, Korea and Russian Far East. Winters across much of Southeast Asia, east to Lesser Sundas and New Guinea. **HABITS AND HABITATS** Uncommon summer breeder and passage migrant. Breeding birds occur in broadleaved evergreen forests, forest edges and woodland. Migrating birds usually occur in small flocks. **TIMING** Mostly Apr–Nov. Spring passage April; autumn passage Sep–Nov. **SITES** Nanling (Guangdong), Wuyuan (Jiangxi), Gutianshan (Zhejiang). **CONSERVATION** Least Concern. Class II protected species.

Eastern Buzzard ▪ *Buteo japonicus* 普通鵟 (Pǔ tōng kuáng) 41-52cm

DESCRIPTION ssp. *japonicas*. Medium-sized, robust-looking raptor. Upperparts dark brown; head to breast pale with fine streaking, contrasting with dark brown of lower flanks, thighs and belly. Juveniles heavily streaked on underparts. In flight, note broad wings and black carpal patches, with flight feathers tipped black. Formerly a race of the **Common Buzzard** (*B. buteo*). **DISTRIBUTION** Breeds E Russia to NE China and Japan. Winters E, NE India, SE China and mainland Southeast Asia .

HABITS AND HABITATS Fairly common passage migrant and winter visitor, occurring in mixed forests, woodland, grassland and farmland; occasionally in inland wetlands. Preys mostly on lizards and rodents caught on the ground. **TIMING** Sep–Apr. **SITES** Suitable habitat across SE China. **CONSERVATION** Least Concern. Class II protected species.

Greater Spotted Eagle ▪ *Clanga clanga* 乌雕 (Wū diāo) 60–71cm

DESCRIPTION Large dark eagle of wetlands; most frequent *Aquila* eagle in SE China. Adults dusky brown with yellow bill. In flight, note broad wings and black flight feathers, with whitish patch to bases of primaries. Young birds (shown) have whitish tips to feathers on much of wings and uppertail-coverts, giving them a spotted appearance; belly streaked

buff. **DISTRIBUTION** Breeds C, NC Europe, eastwards to E Russia and N China. Winters E Sub-Saharan Africa, N Indian subcontinent, SE China and mainland Southeast Asia **HABITS AND HABITATS** Locally uncommon winter visitor. Occurs in open scrub, grassland, paddy fields and coastal wetlands. Mostly hunts small mammals (e.g. rats) and waterbirds, but also known to scavenge on carcasses. **TIMING** Oct–Apr. **SITES** Suitable habitat across SE China (e.g. Futian). **CONSERVATION** Vulnerable. Class II protected species. Threatened by habitat loss in wintering areas and hybridization with closely-related **Lesser Spotted Eagle**.

Eastern Imperial Eagle ■ *Aquila heliaca* 白肩雕 (Bái jiān diāo) 68–84cm

DESCRIPTION Large dark eagle with a pale neck. Adult (shown) plumage mostly dark brown with a distinct golden-buff crown and hind-neck; also white patches on mantle. Young birds tawny-brown with dark flight feathers and tail, heavily streaked buff on upperparts and wings. Larger and longer tailed than **Greater Spotted Eagle**. **DISTRIBUTION** Breeds E Europe, C Asia eastwards to Mongolia and E Siberia. Winters Africa and Middle East, east to N India, S, SE China, rare Southeast Asia. **HABITS AND HABITATS** Rare winter visitor, occurring in open scrub, farmland, and freshwater and coastal wetlands. Usually seen singly. Preys mainly on mammals and large birds. **TIMING** Nov–Mar. **SITE** Poyang Lake (Jiangxi). **CONSERVATION** Vulnerable. Class I protected species. Threatened by habitat loss, persecution and collision with power lines.

Bonelli's Eagle ■ *Aquila fasciata* 白腹隼雕 (Bái fù sǔn diāo) 56–65cm

DESCRIPTION ssp. *fasciata*. Dark raptor with contrasting pale underparts. Adult (shown) head to most of upperparts greyish-brown, contrasting with pale buff mantle (hidden). Underparts dirty white with dark streaking, and barring on lower flanks. In flight, note long wings; also pale scapulars contrasting with dark underwing-coverts. Young birds buff from head to most of underparts. **DISTRIBUTION** Discontinuously from SW Europe, N Africa and Middle East to Indian subcontinent, Myanmar, SE China and Lesser Sundas. In SE China, from Guangdong to Zhejiang. **HABITS AND HABITATS** Uncommon to rare resident of forests, open scrub and woodland in hilly or mountainous areas to over 2,000m, usually where exposed crags are present. Disperses to lowlands in winter, sometimes forage in wetlands. **TIMING** All year round. **SITES** Suitable habitat across SE China (e.g. Wuyishan). **CONSERVATION** Least Concern. Class II protected species.

Pied Falconet ■ *Microhierax melanoleucos* 白腿小隼 (Bái tuǐ xiǎo sǔn) 16cm

DESCRIPTION Smallest raptor in region. Face white with black patch from eye to ear-coverts. Crown and nape to rest of upperparts black, contrasting with white underparts.

White auxiliaries most visible when in flight. In flight, appears compact with pointed wings and longish tail. **DISTRIBUTION** NE India to E China, N Southeast Asia (Vietnam). All recent SE China records from Jiangxi (Wuyuan) and Suichang (Zhejiang). **HABITS AND HABITATS** Locally common resident of broadleaved evergreen forests, forest edges and woodland to 500m. Flies fast with rapid wingbeats. Makes dashing flights to hunt large insects like dragonflies and beetles. **TIMING** All year around. **SITES** Wuyuan (Jiangxi), Suichang (Zhejiang). **CONSERVATION** Least Concern. Class II protected species. Population small but apparently stable.

Common Kestrel ■ *Falco tinnunculus* 红隼 (Hóng sǔn) 30–34cm

DESCRIPTION ssp. *interstinctus*. Small falcon with a long, round-ended tail. Adult male grey headed with a black moustachial stripe; upperparts mostly rufous-brown and finely spotted with black chevrons. Female and young brown headed. In flight, note long, pointed

wings and broad subterminal band on tail. **DISTRIBUTION** Breeds W Europe, Middle East, eastwards to E Russia and much of East Asia. Northern populations winter across Africa, Indian subcontinent, East and Southeast Asia. **HABITS AND HABITATS** Common passage migrant and winter visitor. Occurs in woodland, grassland, open scrub

♀

and farmland. Hovering behaviour while foraging may lead to confusion with unrelated **Black-winged Kite** Preys on small rodents and insects. **TIMING** Most months except Jun. **SITES** Suitable habitat across SE China. **CONSERVATION** Least Concern. Class II protected species.

Swinhoe's Rail ■ *Coturnicops exquisitus* 花田鸡 (Huā tián jī) 13cm

DESCRIPTION Very small, brownish rail. Upperparts brown with broad dark streaking and thin white bars. Flanks brownish with several dark bars; bill and feet olive-yellow. Short tail usually held erect. White secondaries against black primaries visible in flight. **DISTRIBUTION** Breeds S Russian Far East, Japan (Honshu) and NE China. Winters Korea, Japan and SE China. **HABITS AND HABITATS** Rare and local winter visitor to flooded grassland and freshwater marshes. Very secretive and usually only seen when flushed from dense vegetation. Migrating birds may show up in relatively open places, as shown here. **TIMING** Oct–Mar. **SITES** Poyang Lake (Jiangxi), Chongming Dongtan (Shanghai). **CONSERVATION** Vulnerable. Class II protected species. Threatened mainly by habitat loss.

Slaty-legged Crake ■ *Rallina eurizonoides* 白喉斑秧鸡
(Bái hóu bān yāng jī) 25cm

DESCRIPTION ssp. *telmatophila*. Richly coloured rail with dark slaty legs. Head to breast rich orangey-brown, contrasting with dark brown of wings and tail. Belly to undertail-coverts black with thin white barring. Young birds similar to adults, with orangey-brown areas replaced by dark brown. **DISTRIBUTION** Indian subcontinent, Himalayan foothills east to Southeast Asia, S, SE China. In SE China, from Hainan to Fujian. **HABITS AND HABITATS** Mostly summer breeder and passage migrant, occurring in broadleaved evergreen forests, forest edges and woodland at low elevations, usually in dense wet, swampy areas. Also rare winter visitor. Breeding birds mostly vocal at dusk. **TIMING** April–October. **SITES** Nanling (Guangdong); recent records of birds in urban areas in Guangzhou are likely to be migrants. **CONSERVATION** Least Concern. Threatened by habitat loss and trapping.

Brown-cheeked Rail ▪ *Rallus indicus* 普通秧鸡 (Pǔ tōng yāng jī) 27cm

DESCRIPTION Medium-sized rail with a long red bill. Face to breast greyish-blue, contrasting with brown, streaked upperparts due to feathers with dark brown centres. Flanks

to belly black with thin white bars. Bill red, long and thin, recalling a redshank's. Similar looking **Water Rail** (*R. aquaticus*) lacks brown facial band and may occur. **DISTRIBUTION** Breeds Transbaikalia to Russian Far East, NE China, Korea and Japan. Winters S Japan, E, SE China and mainland Southeast Asia. **HABITS AND HABITATS** Uncommon to locally common winter visitor, occurring on freshwater marshes, wet grassland and farmland (e.g. paddy fields). Calls include high-pitched squeaks and nasal squeals. **TIMING** Oct–Apr. **SITES** Dongting Lake (Hunan), Poyang Lake (Jiangxi), Nanhui Dongtan (Shanghai), Xixi National Wetland Park (Zhejiang). **CONSERVATION** Least Concern.

Ruddy-breasted Crake ▪ *Porzana fusca* 红胸田鸡
(Hóng xiōng tián jī) 22cm

DESCRIPTION ssp. *erythrothorax*. Small dark, reddish-brown rail. Upperparts dark brown; head, neck and breast rich chestnut. Lower belly and flanks barred black and

white. Legs bright red. Similar **Slaty-legged** and **Band-bellied Crakes** are much larger. **DISTRIBUTION** Indian subcontinent, Southeast Asia to E Asia as far north as Russian Far East and Japan. Northern populations winter south to SE China and Southeast Asia **HABITS AND HABITATS** Uncommon resident and winter visitor. Occurs in paddy fields, marshes and wet grassland. Difficult to see well due to habit of keeping to thick vegetation, but occasionally ventures to open edges of marshes and scrub to forage. Call is a repeated *teuk* note, usually ending in a shrill, excited trill. **TIMING** All year round. **SITES** Suitable habitat across SE China. **CONSERVATION** Least Concern.

Band-bellied Crake ■ *Porzana paykullii* 斑胁田鸡 (Bān xié tián jī) 21cm

DESCRIPTION Small, richly coloured rail. Face to breast rich orangey-rufous, contrasting with dark greyish-brown of hind-crown, nape and most of upperparts. Belly to undertail-coverts banded with alternating black-and-white bars; white barring on upperwing-coverts. Similar **Ruddy-breasted Crake** is darker and barring starts lower down on belly. DISTRIBUTION Breeds NE China and Russian Far East. Winters Southeast Asia. HABITS AND HABITATS Rare, probably overlooked passage migrant in region. Occurs on freshwater wetlands, wet grassland and farmland (e.g. paddy fields). Hides in dense vegetation, but may occur in open open areas, including urban parks on migration. TIMING May and Oct, during passage periods. SITES No regular sites known; most recent records in region are from Hong Kong. CONSERVATION Near Threatened. Affected by habitat loss at stopover and wintering sites.

Purple Swamphen ■ *Porphyrio porphyrio* 紫水鸡 (Zǐ shuǐ jī) 46cm

DESCRIPTION ssp. *poliocephalus*. Unmistakable large, purplish-blue rail with a bright red frontal shield and bill. Neck to mantle and underparts deep purplish-blue, with greenish-blue patches on breast and scapulars. Back and wings mostly dark brown. DISTRIBUTION Indian subcontinent, Southeast Asia to S, SE, E China. In SE China, mainly from Hainan to Fujian. HABITS AND HABITATS Fairly common resident, occuring on freshwater wetlands, including marshes and well-vegetated ponds and lake sides. Forages mainly for invertebrates; also feeds on plant matter. Vocal, calls a variety of nasal wheezes, squawks and wailing notes. TIMING All year round. SITES Xiamen (Fujian), Gongping Lake (Guangdong). CONSERVATION Least Concern.

Baillon's Crake ■ *Porzana pusilla* 小田鸡 (Xiǎo tián jī) 20cm

DESCRIPTION Small, well-patterned rail. Upperparts rufous-brown speckled black and white; face to upper belly bluish-grey. Flanks and belly to undertail-coverts barred black and white. Young birds (shown) drabber, lack blue-grey of adults; rufous-buff on breast

with fine barring. **DISTRIBUTION** Breeds SW Europe, discontinuously eastwards to E Russia, NE, E China and Japan. Also Africa and Southeast Asia. Northern populations winter Indian subcontinent, Southeast Asia and Australia. **HABITS AND HABITATS** Rare to uncommon passage migrant, probably overlooked. Occurs on freshwater marshes, wet grassland and paddy fields. Very shy, but occasionally ventures out into the open near the edges of marshes and wet grass, especially in the morning to sunbathe. Otherwise mostly seen when flushed. Usually silent. **TIMING** Apr–May, Sep–Oct. **SITES** Suitable habitat across SE China (e.g. Chongming Dongtan, Shanghai). **CONSERVATION** Least Concern.

Juvenile

Brown Crake ■ *Amaurornis akool* 红脚苦恶鸟 (Hóng jiǎo kǔ è niǎo) 26cm

DESCRIPTION ssp. *coccineipes*. Drab brown rail, smaller than similar **White-breasted Waterhen**. Upperparts dark brown from head; underparts bluish-grey; undertail brownish,

contrasting with bright red legs. Throat dirty white. **DISTRIBUTION** Breeds Indian subcontinent, W Myanmar and much of S, SE China, to as far north as Jiangsu. Northern populations undertake short-distance migration. **HABITS AND HABITATS** Uncommon resident of freshwater marshes, wet grassland, inland wetlands and paddy fields, mostly in lowlands. Very shy, often hiding in dense vegetation. Crepuscular; calls a high-pitched, descending trill. **TIMING** All year round. **SITES** Wuyuan, Poyang Lake (Jiangxi), Xixi National Wetland Park (Zhejiang). **CONSERVATION** Least Concern.

White-breasted Waterhen ■ *Amaurornis phoenicurus* 白胸苦恶鸟
(Bái xiōng kǔ è niǎo) 33cm

DESCRIPTION ssp. *phoenicurus*. Medium-sized, slaty-grey and white rail. Upperparts and wings entirely black, contrasting sharply with white face and underparts. Undertail-coverts chestnut. Young birds like adults, but washed brown on face.

DISTRIBUTION Indian subcontinent, Southeast Asia and most of S, E and SE China, east to S Japan. Widespread SE China. **HABITS AND HABITATS** Common resident, occurring in mangroves, freshwater marshes, paddy fields and inland wetlands. Less shy than other rails. Nest a shallow pad in dense vegetation. Commonly heard call is a series of *oo-waks* and monotonous clucks. **TIMING** All year round. **SITES** Suitable habitat across SE China. **CONSERVATION** Least Concern.

Common Moorhen ■ *Gallinula chloropus* 黑水鸡 (Hēi shuǐ jī) 30–36cm

DESCRIPTION ssp. *chloropus*. Duck-like rail with a red frontal shield. Overall plumage slate-grey with dark brown back and wings; white lining along flanks and white undertail-coverts. Yellow-tipped red bill and red frontal shield diagnostic.

DISTRIBUTION Widespread across Americas, Eurasia and Africa. Northern populations migratory, wintering in tropics and subtropics. Widespread SE China. **HABITS AND HABITATS** Common resident and winter visitor, occurring on vegetation-fringed lakes, freshwater marshes, paddy fields and inland wetlands. Usually solitary or in pairs. Calls varied, including loud *krrk* notes when alarmed, and various clucks. **TIMING** All year round. **SITES** Suitable habitat across SE China, (e.g. Xixi National Wetland Park). **CONSERVATION** Least Concern.

left:juvenile

Siberian Crane ■ *Grus leucogeranus* 白鹤 (Bái hè) 140cm

DESCRIPTION Very tall white crane with a red face. Facial skin and legs bright red. Black primaries and primary coverts visible only in flight. Young birds cinnamon-brown,

becoming increasingly mottled with age. **DISTRIBUTION** Breeds Arctic tundra of NW and E Russia (Yakutia). Winters SE China, mainly in Poyang Lake; also Iran. Migrates through NE and E China. **HABITS AND HABITATS** Locally common winter visitor to freshwater marshes, wet grasslands and shallow lakes with vegetated margins, where birds forage for aquatic grasses. Forms small groups of several families during migration. **TIMING** Nov–Feb. **SITES** Poyang Lake (Jiangxi), Chongming Dongtan (Shanghai). **CONSERVATION** Critically Endangered. Class I protected species. Ninety-five per cent of global population of 3,200 birds (2010 estimate) winters in Poyang Lake. The western population may be extinct.

White-naped Crane ■ *Grus vipio* 白枕鹤 (Bái zhěn hè) 127cm

DESCRIPTION Tall dark crane with distinct neck patterns. Plumage mostly grey, contrasting with white neck. Grey extends up neck to nearly meet ear-coverts, forming a white throat-patch. Facial skin red; ear-coverts grey. Wing-coverts pale grey; flight feathers

black. Juveniles washed brown on head and upperparts. **DISTRIBUTION** Breeds E Mongolia, NE China and Russian Far East. Winters mainly S Japan (Kyushu), Korea (DMZ area) and SE China in the lower Yangtze basin. **HABITS AND HABITATS** Locally common winter visitor, occurring on wet grassland, vegetation around lakes and farmland. Family groups congregate to form large flocks, often among other crane species. **TIMING** Nov–Feb. **SITES** Poyang Lake (Jiangxi), Chongming Dongtan (Shanghai). **CONSERVATION** Vulnerable, threatened due to hunting and habitat loss. Class I protected species. About 6,500 individuals remain in the wild.

Hooded Crane

■ *Grus monacha* 白头鹤 (Bái tóu hè) 97cm

DESCRIPTION Small, relatively dark crane. Plumage mostly dark grey with white head and neck. Forehead black, followed by narrow red crown-patch. Flight feathers black. Juveniles pale buff on head; lack crown patterns of adults. **DISTRIBUTION** Breeds Russian Far East and NE China. Winters mainly S Japan (Kyushu) and SE China, in the lower Yangtze basin. **HABITS AND HABITATS** Locally common winter visitor, occurring on wet grassland, vegetation around lakes and farmland. Gregarious; gathers in flocks, sometimes with other cranes, but families keep close together. **TIMING** Nov–Mar. **SITES** Dongting Lake (Hunan), Poyang Lake (Jiangxi), Chongming Dongtan (Shanghai). **CONSERVATION** Vulnerable, threatened due to habitat loss. Class I protected species. Global population estimated at 9,500 individuals.

Grey-headed Lapwing ■ *Vanellus cinereus* 灰头麦鸡

(Huī tóu mài jī) 35cm

DESCRIPTION Distinct plover with a grey head and neck, and a black-tipped yellow bill. Broad dark crescent separates rest of white breast from grey head. Mantle and wing-coverts brown. In flight, black primaries and subterminal bar on uppertail against white backdrop distinctive. **DISTRIBUTION** Breeds NE, E and SE China where widespread. Winters N India to S, SE China and Southeast Asia. **HABITS AND HABITATS** Uncommon winter visitor, passage migrant and summer breeder, occurring in freshwater marshes, on farmland like flooded paddy fields and riverbanks; occasionally also on coastal mudflats and edges of mangroves. Usually in small flocks in winter, but single birds often seen during passage period. **TIMING** Nearly all year round. **SITES** Futian (Guangdong), Poyang Lake (Jiangxi). **CONSERVATION** Least Concern.

Grey Plover ■ *Pluvialis squatarola* 灰斑鸻 (Huī bān héng) 28 cm

DESCRIPTION Similar to **Pacific Golden Plover** but larger, with a stout black bill and whitish brow. Non-breeding birds greyish, with white-and-black spangling on upperparts.

Underparts mostly white. In flight, note white wing-bar across flight feathers and black axillaries contrasting with pale underwing. **DISTRIBUTION** Breeds Arctic tundra of Eurasia and N America. Winters along coasts in tropics and subtropics. Widespread across coastal SE China. **HABITS AND HABITATS** Uncommon winter visitor and passage migrant, occurring on sandy beaches and coastal mudflats. Roosts in dense flocks; forages singly or in well-spaced groups among other plovers for small crabs and molluscs. **TIMING** Sep–May. **SITES** Minjiang Estuary (Fujian), Nanhui Dongtan (Shanghai). **CONSERVATION** Least Concern. As is the case with most shorebirds, habitat loss due to reclamation and hunting are key threats.

Long-billed Plover ■ *Charadrius placidus* 长嘴剑鸻
(Cháng zuǐ jiàn héng) 20cm

DESCRIPTION Medium-sized plover with a 'thin'-looking black bill. In summer head greyish-brown, with a white forehead and brow-patch from behind eye. Ring around neck

mostly black, but augmented by brown patch on breast sides and thus appearing broader there, unlike in similar, but smaller **Little Ringed Plover**. Overall longer bodied than other ringed plovers. **DISTRIBUTION** Breeds C, E and NE China, Russian Far East, Korea, Hokkaido, recently NE India. Winters Himalayan foothills, N Southeast Asia to SW, S, SE and E China. **HABITS AND HABITATS** Uncommon winter visitor, occurring on shingly or pebbly edges of rivers and lakes; also freshwater wetlands and wet fields during migration. Most records are below 600m **TIMING** Mostly Nov–Mar. **SITES** Yanghu Wetlands (Hunan), Wuyuan (Jiangxi). **CONSERVATION** Least Concern. May be affected by pollution and disturbance along rivers.

Kentish Plover ▪ *Charadrius alexandrinus* 环颈鸻 (Huán jǐng héng) 16cm

DESCRIPTION ssp. *nihonensis*, *dealbatus* (shown). Small plover with sandy-brown upperparts, unbroken white collar, narrow black band on breast sides, white forehead and short brow. Race *dealbatus*, or **'White-faced Plover'** is paler and larger headed, with broader white brow and forehead. **DISTRIBUTION** Breeds N Africa and Europe, east to much of E Asia (including coasts of SE China). Northern populations winter in subtropics and tropics. **HABITS AND HABITATS** Uncommon winter visitor and passage migrant. Race *dealbatus* is a summer breeder, occurring on sandy beaches and coastal mudflats in region. Forages by running along beaches, stopping every now and then to pick up invertebrate prey. Sometimes forms mixed flocks with other small shorebirds. **TIMING** All year round. **SITES** Sandy beaches on E and SW Guangdong coast, Minjiang Estuary (Fujian). **CONSERVATION** Least Concern. Habitat loss due to coastal reclamation and hunting are key threats.

Greater Painted Snipe ▪ *Rostratula benghalensis* 彩鹬 (Cǎi yù) 26cm

DESCRIPTION ssp. *benghalensis*. Unmistakable, superficially snipe-like. Female deep chestnut on neck and breast, contrasting with white eye-patch. Broad white band sweeps from breast sides to mantle, separating from dark olive wings. Male similarly patterned but less richly coloured, with buff eye-patch. **DISTRIBUTION** Africa, Indian subcontinent to E, SE China and Southeast Asia. Some populations winter S China and Southeast Asia. **HABITS AND HABITATS** Occurs on freshwater wetlands, wet grassland and farmland as a resident and winter visitor. Also occurs in brackish coastal marshes. Mainly crepuscular, foraging for invertebrates by probing the mud. **TIMING** All year round; most common in autumn months due to influx of migrants. **SITES** Present in suitable habitat in SE China (e.g. Gongping Lake, Guangdong). **CONSERVATION** Least Concern.

Pheasant-tailed Jacana ■ *Hydrophasianus chirurgus* 水雉
(Shuǐ zhì) 32cm + 20cm (tail)

DESCRIPTION Rail-like waterbird with long toes. Breeding adult (shown) unmistakable, with a long black tail and black belly; white head and neck separated from yellow nape by black line. Non-breeding birds less strongly marked and lacks long tail. In flight,

wings mostly white with black feather tips. **DISTRIBUTION** Indian subcontinent, Southeast Asia to S, SE China. Northern populations migrate to Southeast Asia and SE China. **HABITS AND HABITATS** Locally common summer breeder and passage migrant, occurring on well-vegetated wetlands with plants like water lily and water caltrop, on which birds forage. Also on flooded farmland and in mangrove edges during non-breeding period. **TIMING** Breeds May–Aug; passage migrants mostly Oct. **SITES** Zhaoqing (Guangdong), Poyang Lake (Jiangxi), Yanghu Wetlands (Hunan). **CONSERVATION** Least Concern. Loss of freshwater habitats a potential threat.

Asian Dowitcher ■ *Limnodromus semipalmatus* 半蹼鹬 (Bàn pǔ yù) 34cm

DESCRIPTION Superficially resembles **Bar-tailed Godwit**, but smaller and with bulbous-tipped, black bill. Breeding birds rich orange-rufous on head to upper belly, with black barring on flanks. Non-breeding birds greyish-brown on upperparts; paler on underparts, with fine spotting. **DISTRIBUTION** Breeds SW Siberia, N Mongolia east to NE China. Winters Southeast Asia to Australia. **HABITS AND HABITATS** Uncommon passage migrant in region; forages on coastal mudflats and saltpans; rests on sandy beaches and shallow lagoons. Distinctive feeding action involves probing bill into mud repeatedly;

reminiscent of a 'sewing machine'. **TIMING** Spring passage Apr–May; autumn passage late Jul–Sep. **SITES** Minjiang Estuary (Fujian), Chongming Dongtan (Shanghai). **CONSERVATION** Near Threatened. Affected by habitat loss in stopover (e.g. Yellow Sea coast) and wintering areas.

breeding

Black-tailed Godwit ■ *Limosa limosa* 黑尾塍鹬 (Hēi wěi chéng yù) 40 cm

DESCRIPTION ssp. *melanuroides*. Large shorebird with a long, straight bill, not decurved as on **Bar-tailed Godwit**. Bill pinkish at base and tipped black. Breeding birds orange-rufous on head to belly, with black barring on flanks and belly. Non-breeding birds brownish-grey on upperparts, with paler head sides and whitish brow. Underparts whitish. Legs longer than **Bar-tailed Godwit's**. In flight shows prominent white upperwing bars and white band across rump. **DISTRIBUTION** Breeds W Europe, eastwards to NE China and E Russia. Winters throughout Old World subtropics and tropics, eastwards to SW Pacific. Widespread across SE China. **HABITS AND HABITATS** Common winter visitor and passage migrant, occurring mostly in coastal wetlands, especially on mudflats. Forages by probing mud vigorously for worms and molluscs. **TIMING** Aug–May. A few may oversummer. **SITES**
Minjiang Estuary
(Fujian), Futian
(Guangdong), Nanhui
Dongtan (Shanghai).
CONSERVATION
Near Threatened.
Loss of wintering and
stopover habitats in
East and Southeast
Asia a major threat.

breeding

Little Curlew ■ *Numenius minutus* 小杓鹬 (Xiǎo biāo yù) 30 cm

DESCRIPTION Like a miniature **Whimbrel**, but more slender looking and with a shorter, thinner, decurved bill. Back and wings dark brown; head to breast sides washed buff with fine streaking. Note also dark lateral crown-stripe. **DISTRIBUTION** Breeds NE Siberia from Yenisei valley east to Chukotka. Winters across Australia. **HABITS AND HABITATS** Uncommon passage migrant; on passage, occurs on grassland, farmland and even airfields, favouring dryer areas than other shorebirds. Occasionally visits wet grassland and mudflats. **TIMING** Spring passage Apr–early May; autumn passage Sep–Oct. **SITES** Minjiang Estuary (Fujian), Chongming Dongtan (Shanghai). **CONSERVATION** Least Concern.

Eurasian Curlew ■ *Numenius arquata* 白腰杓鹬 (Bái yāo biāo yù) 55cm

DESCRIPTION ssp. *orientalis*. Large shorebird with a very long, decurved bill. Plumage greyish-brown, paler on underparts with fine streaking; lower belly to vent white. White rump and finely barred tail best visible in flight. Young birds darker washed and less streaked on flanks than adults.

DISTRIBUTION Breeds W Europe across Siberia east to NE China. East Asian population winters S Japan, E, SE China to Southeast Asia. **HABITS AND HABITATS** Uncommon to locally common winter visitor and passage migrant, occurring on muddy lake shores, estuaries and coastal mudflats. Most wintering birds arrive in early August, but large flock usually occur after November. A few birds may oversummer in southern parts of region. **TIMING** Mostly Aug–May. **SITES** Minjiang Estuary (Fujian), Chongming Dongtan (Shanghai), Wenzhou Bay (Zhejiang). **CONSERVATION** Near Threatened. Regional population decline documented throughout range.

Eastern Curlew ■ *Numenius madagascariensis* 大杓鹬 (Dà biāo yù) 61cm

DESCRIPTION Large shorebird most similar to **Eurasian Curlew**, but larger and darker, with longer decurved bill. Plumage mostly brown, but darker and richer than **Eurasian**

Curlew's. Neck to belly more boldly streaked; vent washed buff. In flight, rump is buff-brown, not white. **DISTRIBUTION** Breeds E Siberia east to Kamchatka. Winters Southeast Asia east to Australia and New Zealand. **HABITS AND HABITATS** Uncommon passage migrant in region, occurring mostly on coastal mudflats. Aquaculture ponds sometimes used as high-tide roosting sites. **TIMING** Mostly Apr; Aug–Oct during autumn passage. **SITES** Minjiang Estuary (Fujian), Chongming Dongtan (Shanghai). **CONSERVATION** Vulnerable. Threatened mainly by habitat loss in stopover sites (e.g. Yellow Sea).

Nordmann's Greenshank ■ *Tringa guttifer* 小青脚鹬
(Xiǎo qīng jiǎo yù) 30cm

DESCRIPTION Medium-sized shorebird, similar to **Common Greenshank** but paler and stockier, with shorter, yellower tibia. Non-breeding birds (shown) grey on upperparts; underparts white with faint spotting. Unlike in **Common Greenshank**, lightly upturned bill is yellow based and black tipped, not grey. Crown, nape and sides of breast faintly streaked. Toes partly webbed. In flight, underwings white, and feet protrude slightly beyond tail. **DISTRIBUTION** Breeds Sea of Okhotsk coast of E Siberia, Sakhalin Island. Winters Thai-Malay Peninsula, Greater Sundas and Sundarbans. **HABITS AND HABITATS** Uncommon to rare passage migrant, occurring on coastal mudflats. Actively runs on mudflats to catch small crabs. **TIMING** Mostly spring passage Mar–Apr. Also small numbers in autumn and winter. **SITES** Hengsha Island (Shanghai), Wenzhou Bay (Zhejiang). **CONSERVATION** Endangered. Global population less than 1,000 individuals. Threatened by rapid loss and degradation of stopover and winter habitats.

Sanderling ■ *Calidris alba* 三趾滨鹬 (Sān zhǐ bīn yù) 20cm

DESCRIPTION Small pale shorebird; palest of Calidrid sandpipers. Breeding birds orange-rufous on face, breast, mantle and scapulars. Non-breeding birds (shown) pale grey on upperparts; underparts white; dark patch on carpals. Leg and feet black; lacks hind toe, but difficult to see. Young birds darker streaked on crown than adults; mantle and scapulars spangled black and white. In flight, shows broad white wing-bar across black flight feathers. **DISTRIBUTION** Breeds Arctic tundra across N Eurasia and North America. Winters coasts of Americas, Africa, Asia and Australasia. **HABITS AND HABITATS** Common winter visitor and passage migrant, occurring on sandy beaches and sometimes coastal mudflats. Often joins similar-sized waders at roosts. Has a rather peculiar habit of foraging by flipping over stones and debris for small invertebrates. **TIMING** Aug–May. **SITES** Minjiang Estuary (Fujian), Hengsha Island, Nanhui Dongtan (Shanghai). **CONSERVATION** Least Concern.

Red-necked Stint ■ *Calidris ruficollis* 红颈滨鹬 (Hóng jǐng bīn yù) 14cm

DESCRIPTION Small, stocky shorebird with a short black bill. Breeding birds rich orange on head and breast; also rufous on scapulars. Non-breeding birds (shown) brownish-grey on upperparts; underparts white with greyish lateral breast-patch. Smaller and darker than **Sanderling**; bill shorter than very similar **Dunlin's**. Legs and feet black. In flight, note narrow white wing-bar and dark centre to rump and tail. **DISTRIBUTION** Breeds C

Siberia eastwards to Chukotka. Winters coastal Bangladesh, E, SE China, Southeast Asia to Australia, New Zealand. **HABITS AND HABITATS** Common passage migrant, uncommon winter visitor, occurring on coastal mudflats and wetlands, and occasionally on freshwater wetlands. Feeds with rapid pecking action, usually in small groups, among other shorebirds. **TIMING** Most common during passage periods of Apr–May, Aug–Oct. **SITES** Minjiang Estuary (Fujian), Hengsha Island, Nanhui Dongtan (Shanghai). **CONSERVATION** Least Concern.

Long-toed Stint ■ *Calidris subminuta* 长趾滨鹬 (Cháng zhǐ bīn yù) 15cm

DESCRIPTION Small, slim-looking shorebird. In breeding plumage crown chestnut and finely streaked; upperparts and wings chestnut-brown with dark-centred feathers, giving scaly appearance; underparts white with neck sides and breast washed brown

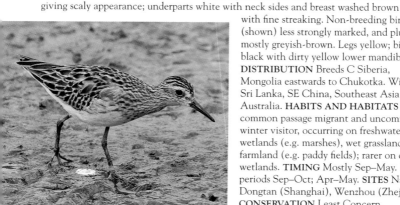

with fine streaking. Non-breeding birds (shown) less strongly marked, and plumage mostly greyish-brown. Legs yellow; bill black with dirty yellow lower mandible. **DISTRIBUTION** Breeds C Siberia, Mongolia eastwards to Chukotka. Winters Sri Lanka, SE China, Southeast Asia to Australia. **HABITS AND HABITATS** Fairly common passage migrant and uncommon winter visitor, occurring on freshwater wetlands (e.g. marshes), wet grassland and farmland (e.g. paddy fields); rarer on coastal wetlands. **TIMING** Mostly Sep–May. Passage periods Sep–Oct; Apr–May. **SITES** Nanhui Dongtan (Shanghai), Wenzhou (Zhejiang). **CONSERVATION** Least Concern.

Dunlin ■ *Calidris alpina* 黑腹滨鹬 (Hēi fù bīn yù) 19cm

DESCRIPTION Medium-sized shorebird, unmistakable in breeding plumage. Breeding adults white on underparts, with fine spotting from face to breast; large black patch from breast to belly distinct. Mantle to scapulars rich chestnut, with dark-centred feathers. Non-breeding birds (shown) greyish-brown on upperparts with darker crown; underparts white with grey wash on breast. Similar to non-breeding **Curlew Sandpiper**, but bill less decurved and legs shorter. **DISTRIBUTION** Breeds Arctic tundra across N Eurasia and North America. Winters

across Americas, Africa, Asia, including S, SE China. **HABITS AND HABITATS** Common winter visitor and passage migrant of coastal mudflats, sandy beaches and coastal wetlands (e.g. marshes); also wet paddy fields. Forms very large flocks. **TIMING** Oct–Apr. **SITES** Suitable habitat across SE China. **CONSERVATION** Least Concern.

Spoon-billed Sandpiper ■ *Eurynorhynchus pygmeus* 勺嘴鹬 (Sháo zuǐ yù) 15cm

DESCRIPTION Small, pale-looking shorebird with a uniquely shaped bill. Breeding birds washed rich orange on head, neck and upper breast, with fine streaking across breast; rest of upperparts brown with pale fringes to feathers. In non-breeding birds, upperparts pale grey and underparts white. Unlike in **Rufous-necked Stint**, head larger, forehead and breast white; note also prominent white brow. Distinctive black, spatulate bill hard to see in profile. **DISTRIBUTION** Breeds mainly NE Russia (Chukotka). Winters coasts of SE China, Bangladesh and mainland Southeast Asia. **HABITS AND HABITATS** Uncommon winter visitor and passage migrant, occurring on coastal mudflats and sandy beaches. Forages on mudflats with backwards-and-forwards and sideways jabbing actions. Usually seen in small groups, pairs or singly. **TIMING** Autumn passage Oct–Nov; spring passage Mar–May; smaller numbers Dec–Feb. **SITES** Minjiang Estuary (Fujian), Leizhou (Guangdong), Hengsha Island (Shanghai). **CONSERVATION** Critically Endangered. Declining due to rapid habitat loss of stopover and wintering sites, and widespread hunting.

breeding

Broad-billed Sandpiper ■ *Limicola falcinellus* 阔嘴鹬 (Kuò zuǐ yù) 17cm

DESCRIPTION Small shorebird with distinct crown patterns. Larger than all stints, with longer and broader black bill, kinked downwards at tip. White brow, lores and narrow white lateral crown-stripe give appearance of a split brow, which is distinctive. Breeding birds brownish from crown to upperparts, with pale-edged feathers on wings; underparts white with dark streaking. Non-breeding birds (shown) grey on upperparts; underparts white with greyish wash on chest. **DISTRIBUTION** Breeds Scandinavia, eastwards to

C, E Siberia. Winters coasts of E Africa, Middle East, eastwards to SE China, Southeast Asia, Australia. **HABITS AND HABITATS** Uncommon to common winter visitor and passage migrant, occurring on coastal mudflats and sandy beaches, and sometimes in paddy fields. Feeds by walking quickly and actively digging into mud for worms. **TIMING** Aug–May, with higher numbers during spring passage Apr–May; autumn passage Aug–Oct. **SITES** Leizhou (Guangdong), Minjiang Estuary (Fujian), Wenzhou Bay (Zhejiang). **CONSERVATION** Least Concern.

Saunders's Gull ■ *Chroicocephalus saundersi* 黑嘴鸥 (Hēi zuǐ ōu) 31cm

DESCRIPTION Small gull with a stubby black bill. Breeding birds black hooded with broken thick white patch around eye. Non-breeding birds pale headed with dark patches joining ear-patches and eye. In flight, note black-and-white bars on outer primaries, and dark patch on underside. First-winter birds brown washed on coverts and tertials. **DISTRIBUTION** Breeds Yellow Sea coast in China (e.g. Liaoning) and Korea. Winters Korea, S Japan to E, SE China, N Vietnam. **HABITS AND HABITATS** Rare to locally

common winter visitor of coastal mudflats. Flies buoyantly, suddenly diving to catch crabs in coastal areas. Most adults leave wintering grounds in March, but young birds may stay

longer. **TIMING** Oct–Apr. **SITES** Minjiang Estuary, Fuqing/Putian coast (Fujian), Wenzhou Bay (Zhejiang). **CONSERVATION** Vulnerable. Class II protected species. Threatened by loss of mudflat habitats, spread of invasive smooth cordgrass and disturbance by fishermen.

breeding

Relict Gull ■ *Ichthyaetus relictus* 遗鸥 (Yí ōu) 42cm

DESCRIPTION Medium-sized gull with a thick-necked appearance. Adults light grey on mantle and wings; note black-and-white wing-tips when wings are folded. Breeding birds black hooded, contrasting with thick pale eyelid and reddish bill. Non-breeding birds (shown) white headed; finely streaked on nape; bill black. Immatures white headed; tertials dark; black on outer primaries; thin black terminal bar on upper tail.
DISTRIBUTION Breeds inland lakes C Asia, Mongolia to N China. Winters mostly Bohai Gulf, south to E and SE China coast. **HABITS AND HABITATS** Uncommon to rare winter visitor, occurring mostly on coastal mudflats. When foraging, makes short flights with sudden drops to catch crabs on mudflats. **TIMING** Nov–Mar. **SITES** Minjiang Estuary (Fujian), Chongming Dongtan (Shanghai). **CONSERVATION** Vulnerable. Class I protected species. Threatened by reclamation and disturbance from shellfish collectors and fishermen.

Black-tailed Gull ■ *Larus crassirostris* 黑尾鸥 (Hēi wěi ōu) 46cm

DESCRIPTION Medium-sized gull. Adult dark grey on mantle; bill yellow with black tip; thick black band on tail. First-winter birds mostly brown with whitish forehead; bill pink with black tip. Second-winter birds greyish on mantle; bill pale yellowish. **DISTRIBUTION** Breeds Russian Far East, Yellow Sea coast (including China and Korea), Japan and E China. Winters coasts of Japan, E and SE China, including Taiwan. **HABITS AND HABITATS** Common summer breeder on remote, rocky islands off Fujian, Zhejiang coasts. Also occurs in various coastal habitats including harbours and mudflats in winter, where birds forage for small fish and crustacean, also scavenges. **TIMING** All year round, most numerous Nov–Mar, and especially spring passage in Mar. **SITES** Many coastal areas from Shanghai to Fujian (e.g. Minjiang). **CONSERVATION** Least Concern.

first winter

adult

Heuglin's Gull ▪ *Larus heuglini* 乌灰银鸥 (Wū huī yín ōu) 60cm

DESCRIPTION Large, strongly built gull with a long-winged appearance. Adults dark grey on mantle, nape white but heavily streaked in autumn–winter; bill yellow with red spot on mandible base; legs invariably coloured from yellow to pink. First-winter birds dark brownish-grey on mantle and upperwings; some brown streaks on lower breast; bill pinkish with black tip; thick black band on tail. Second-winter birds greyish on mantle; less streaked. Some authorities treat it as a race of **Lesser Black-backed Gull** (*L. fuscus*). **DISTRIBUTION** Breeds Arctic coasts of NW Russia (Kola) to C Siberia (Taimyr).

Winters south to coasts of Africa, Indian subcontinent, Southeast Asia and across E Asia. **HABITS AND HABITATS** Common winter visitor to coastal habitats including mudflats, beaches and harbours; sometimes in coastal wetlands. Forages for molluscs and crabs; also scavenges. **TIMING** Nov–Apr. **SITES** Many coastal sites across SE China. **CONSERVATION** Least Concern.

adult

second winter

Vega Gull ▪ *Larus vegae* 西伯利亚银鸥 (Xī bó lì yà yín ōu) 62cm

DESCRIPTION Largest 'herring gull' in region. Adults mid-grey on upperparts, heavily streaked on nape to breast sides; legs deep pink. First-winter birds similar to those of **Heuglin's Gull**, but heavier, rounder on head, breast deeper and bill thicker; also pale 'windows' more obvious on inner primaries. Lower breast and belly usually blotched

brown. Pale areas of mantle and coverts contrast more strongly with dark subterminal bar. Second-winter birds greyer on mantle. Some authorities treat it as a race of **Herring Gull** (*L. argentatus*). **DISTRIBUTION** Breeds N Siberia (Taimyr), east to Chukotka. Winters coasts of Korea, Japan, E, SE China, including Taiwan. **HABITS AND HABITATS** Common winter visitor, increasingly rare further south, in coastal habitats including mudflats, estuaries and coastal wetlands. Like other gulls, forages for fish and crustacean, also frequently scavenging. **TIMING** Nov–Apr. **SITES** Sporadically reported from across region, but probably widespread on SE China coasts. **CONSERVATION** Least Concern.

Greater Crested Tern ▪ *Thalasseus bergii* 大凤头燕鸥
(Dà fèng tóu yàn ōu) 47cm

DESCRIPTION ssp. *cristatus*. Large sea tern. Resembles **Lesser Crested Tern**, but larger and stockier; bill light yellow, and more robust. Uppersides darker grey. **DISTRIBUTION** Breeds across tropical and subtropical Indian and Pacific Oceans, including islets off SE China coast (e.g. Yushan, Jiushan). **HABITS AND HABITATS** Fairly common summer visitor and passage migrant in coastal waters; often rests on coastal mudflats, rocky outcrops and floating objects. Breeds on offshore islands in large colonies. Flies rapidly with deep, slow wingbeats. Hunts by diving and plunging into the water. **TIMING** Apr–Oct. **SITES** Minjiang Estuary (Fujian), Jiushan, Yushan, Wuzhishan Islands (Zhejiang). **CONSERVATION** Least Concern. Breeding colonies often disturbed by fishermen.

Chinese Crested Tern ▪ *Thalasseus bernsteini* 中华凤头燕鸥
(Zhōng huá fèng tóu yàn ōu) 40cm

DESCRIPTION Medium-sized pale tern with a black crest. Diagnostic black-tipped yellow bill absent in similar **Greater Crested Tern**. Cap black during breeding season; forecrown white in non-breeding birds. Upperparts pale grey, underparts white. **DISTRIBUTION** Breeds islets off E China. Winters in seas around Southeast Asia (e.g. Wallacea). **HABITS AND HABITATS** Summer breeder on rocky islets off Chinese coast, usually among colonies of **Greater Crested Tern**. Often rests on coastal flats during low tide. **TIMING** Apr–Oct. **SITES** Minjiang Estuary (Fujian), Wuzhishan, Jiushan (Zhejiang). **CONSERVATION** Critically Endangered. Class II protected species. Global population less than 50 mature individuals. Of three breeding groups known, the highest consists of 19 adults in Jiushan in 2013. Threatened by habitat disturbance, hybridization, egg collecting and damage to nests by typhoons.

breeding

Little Tern ■ *Sternula albifrons* 白额燕鸥 (Bái é yàn ōu) 25cm

DESCRIPTION ssp. *sinensis*. Smallest tern in region. Breeding birds have a black-tipped yellow bill, and a white forehead extending to over eye; black from eye-stripe to hindneck. Upperparts grey with outer primaries darker; underparts white. Non-breeding birds

black billed, with more white on crown. **DISTRIBUTION** Breeds discontinuously W Europe eastwards to East and Southeast Asia, Australia. Winters coastal areas across tropics and subtropics. **HABITS AND HABITATS** Fairly summer breeder and passage migrant, occurring on coastal wetlands, tidal creeks and saltpans. Breeds coasts and riverbanks, usually on sandy, shingly areas. Hovers and plunge-dives while foraging. Usually in small flocks. **TIMING** Apr–Sep. Passage birds present Apr–May; Aug–Sep. Breeding birds mostly seen May–Aug. **SITES** Minjiang Estuary (Fujian), Haifeng (Guangdong). **CONSERVATION** Least Concern. May be susceptible to loss of breeding habitat in coastal areas.

Gull-billed Tern ■ *Gelochelidon nilotica* 鸥嘴噪鸥 (Ōu zuǐ zào ōu) 39cm

DESCRIPTION ssp. *affinis*. Medium-sized tern with a robust black bill. Breeding birds have a black cap that contrasts sharply with white cheeks and underparts; upperparts pale grey, with dark-tipped primaries. Non-breeding birds black on ear-coverts and in front of eye, giving masked appearance. **DISTRIBUTION** Breeds discontinuously North

America, SW Europe and N Africa, to Mongolia, NE and E China. Winters south to coasts of South America, Africa, Indian subcontinent, S China and Southeast Asia. **HABITS AND HABITATS** Uncommon summer breeder and passage migrant. Occurs on coastal mudflats and estuaries, in coastal waters and on freshwater lakes. Hovers and dives for fish; also preys on insects. **TIMING** Mar–Oct. Mostly late Mar–early May; Aug–Oct. **SITES** Minjiang Estuary (Fujian), Nanhui Dongtan (Shanghai). **CONSERVATION** Least Concern.

Whiskered Tern ■ *Chlidonias hybrida* 须浮鸥 (Xū fú ōu) 26cm

DESCRIPTION ssp. *hybrida*. Small dark tern. Breeding birds black capped; whitish cheeks contrast with grey underparts, giving whiskered appearance; bill red. Non-breeding birds (shown) similar to smaller **White-winged Tern**, black on mask and hind-crown, while forecrown and lores white. Ear-coverts, hind-crown and nape blackish. Note also dark reddish legs and feet. **DISTRIBUTION** Breeds W Europe, Middle East and parts of C Asia eastwards to NE China and Russian Far East. Winters Africa, Indian subcontinent and Southeast Asia, east to Australia. **HABITS AND HABITATS** Common summer breeder, passage migrant and winter visitor. Occurs mostly on freshwater lakes, rivers, wet grassland and paddy fields. Dips into water to pick prey from surface. **TIMING** Nearly all year round, but most common during passage Apr–May, Sep–Oct. **SITES** Yanghu Wetlands (Hunan), Poyang Lake, Nanshan Wetlands (Jiangxi). **CONSERVATION** Least Concern.

Barred Cuckoo-Dove ■ *Macropygia unchall* 斑尾鹃鸠
(Bān wěi juān jiū) 39cm

DESCRIPTION ssp. *minor*. Large pigeon with a long tail. Head and neck bluish-grey; rest of upperparts brown-black, finely barred rufous-brown. Underparts vinous-brown, with breast and flanks barred black. Greenish metallic sheen on nape and neck distinct; less so on females. **DISTRIBUTION** Himalayan foothills, S and SE China to Southeast Asia. In SE China, from Hainan to Zhejiang. **HABITS AND HABITATS** Uncommon resident of montane broadleaved evergreen forests, usually from 500 to 1,900m, sometimes descending to lower hills. Usually seen feeding quietly at fruiting trees; sometimes descends to ground to ingest soil. Calls a low, repeated *oOOo*. **TIMING** All year round. **SITES** Fuzhou Forest Park (Fujian), , Wuyishan (Jiangxi), Jianfengling (Hainan). **CONSERVATION** Least Concern. Class II protected species.

Thick-billed Green Pigeon ■ *Treron curvirostra* 厚嘴绿鸠
(Hòu zuǐ lǜ jiū) 27cm

DESCRIPTION ssp. *hainana*. Colourful forest pigeon with a large, bluish-green patch around eye and red bill base. Males bluish-grey on head, intergrading into pale green of

neck and underparts. Wings and back rich maroon. Females (shown) mostly green. **DISTRIBUTION** Himalayan foothills, E India to S China; also Southeast Asia. In SE China, mostly Hainan, but stragglers have reached Hong Kong. **HABITS AND HABITATS** Uncommon resident of broadleaved evergreen forests, mostly in the lowlands. Often feeds in small flocks at fruiting trees; known to wander widely. Calls a soft, nasal *ooo-ii-ooo*, often repeated. **TIMING** All year round. **SITE** Bawangling (Hainan). **CONSERVATION** Least Concern. Class II protected species. Hunting and habitat loss are key threats.

White-bellied Green Pigeon ■ *Treron sieboldii* 红翅绿鸠
(Hóng chì lǜ jiū) 33cm

DESCRIPTION ssp. *sieboldii* (shown), *murielae* (Hainan, S Guangdong). Medium-sized green pigeon. Male's head to breast yellowish-green; mantle grey; back and wings green

with maroon scapular patch. Belly, vent and undertail-coverts white with bold green bars. Female (shown) similar to male but scapulars green. **DISTRIBUTION** N Southeast Asia (Vietnam), C, S and SE China, Taiwan and Japan. **HABITS AND HABITATS** Uncommon resident and winter visitor occurring in broadleaved evergreen forests, including secondary forests, from lowlands to about 2,000m. Feeds in fruiting trees in small groups, elsewhere has been observed to gather at coastal site to drink seawater. Flight fast with heavy flapping. **TIMING** All year round. **SITES** Jianfengling (Hainan), Jiulianshan (Jiangxi), Nanhui Dongtan (Shanghai). **CONSERVATION** Least Concern. Class II protected species.

Mountain Imperial Pigeon ■ *Ducula badia* 山皇鸠
(Shān huáng jiū) 45cm

DESCRIPTION ssp. *griseicapilla*. Largest pigeon in region. Head and underparts mostly bluish-grey, with neck tinged vinous-pink. Back and wings maroon-brown, not green as

in **Green Imperial Pigeon**; flight feathers black. Note also pale irises, and red orbital skin and bill base (ssp. *badia* from Southeast Asia shown). **DISTRIBUTION** Himalayan foothills, S India, S, SE China and Southeast Asia. In SE China, only Hainan. **HABITS AND HABITATS** Uncommon resident of broadleaved evergreen forests, mostly above 500m. Often disperses into lowlands to feed at fruiting trees, sometimes even in mangroves and agricultural areas. Call a deep, guttural *croo*. **TIMING** All year round. **SITES** Jianfengling, Bawangling (Hainan). **CONSERVATION** Least Concern. Class II protected species. Commonly hunted for food in some parts of China.

Greater Coucal ■ *Centropus sinensis* 褐翅鸦鹃 (Hè chì yā juān) 52cm

DESCRIPTION ssp. *sinensis*. Larger of two coucals; more robust and with longer tail. Entirely glossy black except for chestnut wings and mantle. Resembles **Lesser Coucal**, but lacks streaking on back, and wings are chestnut, appearing 'cleaner'. Juveniles similar

to adults, but with fine black barring on wings. **DISTRIBUTION** Indian subcontinent, S China and Southeast Asia. Widespread in SE China, from Hainan to Zhejiang. **HABITS AND HABITATS** Common resident. Occurs in forest edges, mangroves, scrub and well-vegetated farmland. Often forages walking on the ground, or in low shrubs, for large insects and lizards. Calls a long, descending series of *boop* notes. **TIMING** All year round. **SITES** Suitable habitat across SE China. **CONSERVATION** Least Concern. Class II protected species. Hunted for use in traditional medicine.

Indian Cuckoo ▪ *Cuculus micropterus*
四声杜鹃 (Sì shēng dù juān) 33cm

DESCRIPTION ssp. *micropterus*. Medium-sized cuckoo similar to **Himalayan Cuckoo**. Head grey; back, mantle and wings greyish-brown, unlike grey of **Himalayan Cuckoo**. White underparts barred with widely spaced black bands from breast to vent. Grey tail ends with a prominent broad black subterminal band, tipped white. **DISTRIBUTION** Breeds Indian subcontinent, Southeast Asia, SE, E and NE China to Russian Far East and Korea. Northern populations winter Southeast Asia, also parts of S and C India. **HABITS AND HABITATS** Fairly common summer breeder over most of SE China, also winter visitor to Hainan. Occurs in broadleaved evergreen and deciduous forests, mostly in lowlands and hills. Brood parasite of **Ashy** and **Black Drongo**. Calls a four note *ko-ko-ko-ku*, the last note lowest. **TIMING** Mostly Apr–Aug; All year round in Hainan, but commoner Oct–Mar. **SITES** Suitable habitat across SE China. **CONSERVATION** Least Concern.

Himalayan Cuckoo ▪ *Cuculus saturatus* 中杜鹃 (Zhōng dù juān) 33cm

DESCRIPTION Medium-sized cuckoo very similar to **Oriental** and **Eurasian Cuckoos**. Head paler grey than hind-neck and most of upperparts, unlike more uniform grey of

Eurasian Cuckoo. White underparts barred with black bands (thicker in **Eurasian Cuckoo**), extending to buffy-yellowish (not white) vent. Hepatic morph birds (shown) are barred rufous and dark brown, with brown barring on upperparts. **DISTRIBUTION** Breeds Himalayan foothills to much of S, C and SE China, including Taiwan. Winters mainland Southeast Asia, east to New Guinea. **HABITS AND HABITATS** Fairly common summer breeder and passage migrant. Occurs in broadleaved evergreen and mixed forests from hilly elevations to 1,900m. Brood parasite of *Phylloscopus* and *Seicercus* warblers. Calls a 4–5-note *KU-koo-koo-koo*, the first note highest. **TIMING** Apr–Oct. **SITES** Badagongshan (Hunan), Wuyishan (Jiangxi). **CONSERVATION** Least Concern.

Mountain Scops Owl ■ *Otus spilocephalus* 黄嘴角鸮
(Huáng zuǐ jiǎo xiāo) 20cm

DESCRIPTION ssp. *latouchi*. Richly coloured scops
owl with short ear-tufts. Crown to rest of upperparts
rufous-brown with fine mottling; distinct white bar
across scapulars. Underparts buff-brown, unlike
white of rufous-morph **Oriental Scops Owl**, and
finely streaked; pale stripe from breast to belly.
Eyes yellow (ssp. *hambroecki* from Taiwan shown).
DISTRIBUTION Himalayas, NE India, Southeast
Asia to SW, S, SE China, including Taiwan.
Widespread in SE China. **HABITS AND HABITATS**
Uncommon resident of broadleaved and mixed
evergreen forests mostly above 1,000m. Calls a high-
pitched *wu-wu* at regular intervals. Forages below
canopy to understorey, hunting for large insects and
small mammals. **TIMING** All year round. **SITES**
Heishiding, Nanling (Guangdong), Jianfengling
(Hainan). **CONSERVATION** Least Concern. Class II
protected species.

Oriental Scops Owl ■ *Otus sunia* 红角鸮 (Hóng jiǎo xiāo) 19cm

DESCRIPTION ssp. *stictonotus* (shown, grey morph), *malayanus* (Hainan, S Guangdong).
Small scops owl with prominent ear-tufts; occurs as grey and rufous morphs. Grey-morph
birds greyish-brown on upperparts with fine mottling, black spotting on crown and pale
bar across scapulars; underparts finely vermiculated grey
with brown streaking. Rufous-morph birds rich rufous-
brown on upperparts; whitish on underparts with dark
vermiculation and streaking. Eyes yellow in all morphs.
DISTRIBUTION Breeds Indian subcontinent, Himalayas,
Southeast Asia to NE China, Russian Far East, Japan.
Widespread in SE China. Northern populations winter
Southeast Asia. **HABITS AND HABITATS** Fairly
common resident and migrant across region, occurring in
broadleaved evergreen and mixed forests, forest edges and
woodland to 900m, including parkland during migration.
Calls repeated 3 note series *Kwut-ko-kor*. **TIMING** All
year round. Most common during autumn passage in Oct.
SITES Nanling (Guangdong), Jianfengling (Hainan),
Longqishan (Fujian). **CONSERVATION** Least Concern.
Class II protected species. Like most other owls, trapped
for traditional medicine.

Brown Fish Owl ■ *Ketupa zeylonensis*
褐渔鸮 (Hè yú xiāo) 53cm

DESCRIPTION ssp. *orientalis*. Large owl with ear-tufts and unfeathered legs. Upperparts dull brown with bold streaking; sometimes white patches on scapulars; thick barring on flight feathers. Underparts yellowish-buff with dark brown streaking. Throat whitish. Eyes yellow. **DISTRIBUTION** Indian subcontinent, Southeast Asia to S China. In SE China, Hainan to Guangdong. **HABITS AND HABITATS** Uncommon resident of broadleaved evergreen forests and woodland from lowlands to 1,500m, usually near unpolluted streams and reservoirs. Nocturnal; sometimes seen at daytime roosts. Feeds on fish and snakes. **TIMING** All year round. **SITES** Under-recorded in region; should be searched for in suitable habitat. **CONSERVATION** Least Concern. Class II protected species. Like most large owls, may be susceptible to trapping in S China.

Eurasian Eagle-Owl ■ *Bubo bubo* 雕鸮
(Diāo xiāo) 60–70cm

DESCRIPTION ssp. *swinhoei*. Largest owl in region, with prominent, horn-like ear-tufts. Upperparts and wings brownish with heavy dark brown mottling. Well-defined facial disc greyish-brown and rimmed black; eyes orange. Underparts sandy-brown and boldly streaked from throat to breast; finely barred on flanks and belly. Similar **Tawny Fish Owl** streaked heavily across most of buffy-brown underparts. **DISTRIBUTION** Widespread across Eurasia, from W Europe, Middle East, C Asia eastwards to E Russia, Japan and E, SE China. **HABITS AND HABITATS** Uncommon to rare resident of forest edges, woodland and farmland in hilly and mountainous regions, especially areas with open rocky outcrops. Calls a two-syllable *wu-huu*. **TIMING** All year round. **SITES** Few records in SE China although regular in Hong Kong, including areas on border with Shenzhen. **CONSERVATION** Least Concern. Class II protected species.

Brown Wood Owl ■ *Strix leptogrammica* 褐林鸮 (Hè lín xiāo) 50cm

DESCRIPTION ssp. *ticehursti*. Large, earless owl with dark 'eye shadows'. Well-defined facial disc with a pale brow; black around eyes graduating to buff, and rimmed black, giving a spectacled appearance. Upperparts and wings rich brown, with pale buff barring on wings. Underparts buff, darker towards chest and finely barred dark brown. **DISTRIBUTION** Indian subcontinent, Southeast Asia to SW, S, SE China, including Taiwan. Widespread SE China. **HABITS AND HABITATS** Uncommon resident of broadleaved evergreen and mixed forests in hilly areas, and sometimes forest edges and woodland, from 200m to 1,500m. Calls a *hu-hu-hu-hoo-oo*, the last note longest. **TIMING** All year round. **SITES** Heishiding (Guangdong), Jianfengling (Hainan), Jiulianshan (Jiangxi), Jiulongshan (Zhejiang). **CONSERVATION** Least Concern. Class II protected species. Recently colonized Hong Kong, possibly from source populations in Guangdong.

Northern Boobook ■ *Ninox japonica* 北鹰鸮 (Běi yīng xiāo) 30cm

DESCRIPTION ssp. *japonica*. Medium-sized, earless owl with a hawk-like face. Rounded head dark greyish-brown, with a poorly defined facial mask and pale buff on forehead and chin sides. Upperparts rich chocolate-brown; underparts white with several rows of teardrop-shaped streaks from throat to belly. **DISTRIBUTION** Breeds C, SE China to NE China, Russian Far East, Japan. Most populations winter in Southeast Asia (except Taiwan, Ryukyus). **HABITS AND HABITATS** Uncommon summer breeder and passage migrant, occurring in broadleaved evergreen and mixed forests, forest edges and woodland to about 1,500m. During migration can occur even in urban parklands. Calls a two-note *wuh-wuh* of similar pitch. **TIMING** Mar–Nov. **SITES** Present in sites with suitable habitats (e.g. Wuyuan), has been recorded on migration in Sun Yat-sen University (Guangdong) and Nanhui Dongtan (Shanghai) **CONSERVATION** Least Concern. Class II protected species.

Grey Nightjar ▪ *Caprimulgus jotaka* 普通夜鷹 (Pǔ tōng yè yīng) 28cm

DESCRIPTION ssp. *jotaka*. Greyish-brown nightjar with buff spots on wings. Back heavily vermiculated, with pale buff scapulars. Larger and darker than **Savannah Nightjar**. Crown darker than in **Large-tailed Nightjar** (Hainan), and white throat-patch smaller. **DISTRIBUTION** Widespread Indian subcontinent east to Russian Far East, Japan, NE, E and SE China, also Palau. Also mainland Southeast Asia. Northern populations winter

Southeast Asia. **HABITS AND HABITATS** Fairly common summer breeder and passage migrant. Occurs in broadleaved evergreen forests and forest edges from hills to 1,700m; also urban parks on migration. Feeds by sallying quietly over forests at dusk. Calls a repeated, rapid series of *chupchupchup*. **TIMING** Mar–Nov. **SITES** Suitable habitat across SE China. **CONSERVATION** Least Concern.

Pacific Swift ▪ *Apus pacificus* 白腰雨燕 (Bái yāo yǔ yàn) 15cm

DESCRIPTION ssp. *pacificus* (shown), *kanoi*. Slender-looking swift with a deeply forked tail and white rump-patch. Overall blackish-brown, underparts scaled buff and pale throat. Resident **House Swift** smaller, darker and with tail slightly notched, not forked. **DISTRIBUTION** Breeds C Asia to Russian Far East, NE and E China, and Japan. Most populations winter

Southeast Asia to Australia. Widespread SE China. **HABITS AND HABITATS** Uncommon summer breeder (ssp. *kanoi* only), passage migrant and winter visitor (ssp. *pacificus*). Nests in coastal cliffs and man-made structures, foraging over diverse habitats. Flight erratic, frequently turning. Forms mixed groups with other swifts while feeding. **TIMING** All year round, but commoner Apr–Aug. **SITES** Suitable habitat across SE China. **CONSERVATION** Least Concern.

underparts   *upperparts*

Red-headed Trogon ■ *Harpactes erythrocephalus* 红头咬鹃
(Hóng tóu yāo juān) 33cm

DESCRIPTION ssp. *yamakensis*, *hainanus* (Hainan, shown). Only trogon in region.
Head and neck bright red, with bluish-violet facial skin; rest of upperparts cinnamon-
brown. Underparts brighter red than head, demarcated by white
breast-band. Female rufous-brown on head and upper breast, not
red. **DISTRIBUTION** Himalayan foothills,
NE India, S China and Southeast Asia.
Widespread SE China from Hainan to
Zhejiang. **HABITS AND HABITATS** Fairly
common resident of broadleaved evergreen
forests, from 400 to 1,500m. Forages quietly
for large insects in forest canopy, sometimes
joins mixed flocks. Calls a melancholy series
of 5–6 *tiaup* notes. **TIMING** All year round.
SITES Fuzhou Forest Park (Fujian), Jiulianshan
(Jiangxi), Jianfengling, Bawangling (Hainan).
CONSERVATION Least Concern. Likely
threatened by habitat loss and degradation.

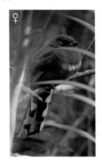

Common Dollarbird ■ *Eurystomus orientalis* 三宝鸟 (Sān bāo niǎo) 30cm

DESCRIPTION ssp. *calonyx*. Dark
greenish-blue roller with a robust red
bill. Plumage (including wings) dark
glossy greenish-blue, paler on belly,
darkest on head. Pale bluish patches
on flight feathers visible only in flight.
DISTRIBUTION Breeds E India to E
Asia north to Russian Far East; also
Southeast Asia east to Australia and W
Pacific Islands. Northern populations
migrate to winter across S China
and Southeast Asia. **HABITS AND
HABITATS** Common summer breeder
and passage migrant. Occurs at edges of
broadleaved evergreen and mixed forests,
especially on adjacent farmland and in
open country. Dependent on natural tree
cavities for nesting. Calls a harsh series of
cough-like cackles. **TIMING** Mostly Apr–
Oct. **SITES** Suitable habitat across SE
China. **CONSERVATION** Least Concern.

Black-capped Kingfisher ■ *Halcyon pileata* 蓝翡翠 (Lán fěi cuì) 29cm

DESCRIPTION Only kingfisher with a black head. Neck to breast white; rest of underparts rich orange. Rest of upperparts including wings deep blue. White patches on flight feathers

visible only in flight. **DISTRIBUTION** Breeds Indian subcontinent, across S and E China, to Korea and Japan. Northern populations winter Southeast Asia and S India. Widespread SE China. **HABITS AND HABITATS** Uncommon summer breeder, passage migrant and winter visitor. Occurs in woodland abutting rivers, inland wetlands, farmland (including paddy fields) and mangroves. Mostly in lowlands to 600m. Calls a shrill, staccato *krk*, especially in flight. **TIMING** All year round. **SITES** Futian (Guangdong), Wuyuan (Jiangxi), Nanhui Dongtan (Shanghai). **CONSERVATION** Least Concern.

Blyth's Kingfisher ■ *Alcedo hercules* 斑头大翠鸟 (Bān tóu dà cuì niǎo) 23cm

DESCRIPTION Medium-sized dark kingfisher. Similar to **Common Kingfisher** but darker, larger and lacks rufous patch on ear-coverts. Upperparts dark blue-green, with white patch on neck sides. Underparts and loral patch orange-rufous. Bill black in males, red-based in females. **DISTRIBUTION** NE India, mainland Southeast Asia to SW, S, SE China. In SE China, from Hainan to Fujian. **HABITS AND HABITATS** Uncommon to locally common resident of broadleaved evergreen forests from hills to 900m, usually near clear streams.

Generally seen perched on rocks or overhanging branches. Flight heavier than **Common Kingfisher's**. **TIMING** All year round. **SITES** Longqishan

(Fujian), Chebaling (Guangdong), Bawangling (Hainan), Jiulianshan (Jiangxi). **CONSERVATION** Near Threatened. Affected by disturbance and pollution of forest streams.

Crested Kingfisher ■ *Megaceryle lugubris* 冠鱼狗 (Guān yú gǒu) 41cm

DESCRIPTION ssp. *guttulata*. Very large pied kingfisher with a striking crest. Crest and much of upperparts finely striped black and white. Neck sides and underparts white, broken by dark stripe across cheeks. Underwing-

coverts rufous in females (shown).
DISTRIBUTION Himalayas east to N
Southeast Asia, across most of E China and
Japan. Widespread SE China.
HABITS AND HABITATS Locally
common to uncommon resident of
broadleaved and mixed evergreen forests
from hills to about 1,000m, usually near
clear, fast-flowing rivers. Perches on rocks,
overhanging branches and overhead wires,
from where it hunts. Habit of bobbing head
and body when hunting. **TIMING** All year
round. **SITES** Chebaling (Guangdong),
Wuyuan (Jiangxi), Gutianshan (Zhejiang).
CONSERVATION Least Concern.

Blue-throated Bee-eater

■ *Merops viridis* 蓝喉蜂虎
(Lán hóu fēng hǔ) 22cm

DESCRIPTION Greenish bee-eater with
a dull blue throat. Head and back rich
chestnut with black eye-stripe. Upperparts
dark green and underparts paler green; tail
and rump blue. Juveniles duller and lack
tail streamers. **DISTRIBUTION** Breeds S,
SE China and Southeast Asia, eastwards to
Philippines. SE China populations winter
Southeast Asia. In SE China, from Hainan to
Fujian. **HABITS AND HABITATS** Common
summer breeder and passage migrant. Occurs
in edges of broadleaved evergreen forests,
woodland and farmland, from lowlands to
500m. Excavates burrows in sandy banks to
nest, often colonially. Hunts for large flying
insects, caught on the wing. **TIMING** Mostly
Apr–Oct. **SITES** Xiamen University (Fujian),
Wuyuan (Jiangxi), Nanling (Guangdong).
CONSERVATION Least Concern.

Great Barbet ■ *Megalaima virens* 大拟啄木鸟 (Dà nǐ zhuó mù niǎo) 34cm

DESCRIPTION ssp. *virens*. Distinctive 'large-headed' barbet with a bluish-black head and a robust, pale bill. Upper back washed rufous; rest of upperparts, including tail, green.

Underparts pale yellow with bold streaking. **DISTRIBUTION** Himalayan foothills, NE India, S China, and mainland Southeast Asia. Widespread in SE China, from Guangdong to N Zhejiang. **HABITS AND HABITATS** Common resident in broadleaved evergreen and mixed forests from hills to about 1,900m. Forages for fruits high up in canopy; often difficult to see. Flight undulating, similar to that of woodpeckers. Calls a monotonous series of *kee-you*, often in duet with mate. **TIMING** All year round. **SITES** Nanling (Guangdong), Jiulianshan (Jiangxi), Tianmushan (Zhejiang). **CONSERVATION** Least Concern.

Chinese Barbet ■ *Megalaima faber* 黑眉拟啄木鸟
(Hēi méi nǐ zhuó mù niǎo) 21cm

DESCRIPTION ssp. *sini*, *faber* (Hainan; shown). Small, all-green barbet with colourful facial patterns. Forecrown black and hind-crown red; cheeks to neck sides blue; throat

yellow with small red patch on bib. Recently split from **Black-browed Barbet** (*M. oorti*). **DISTRIBUTION** SE China. In SE China, from Hainan to Jiangxi, Fujian. Endemic to China. **HABITS AND HABITATS** Common resident of broadleaved evergreen and mixed forests from hills to 1,500m. Forages in canopy for fruits (e.g. figs) and occasionally insects. Calls a monotonous *choo-Chok-Chok-Chok-Chorok*, usually from a high perch. **TIMING** All year round. **SITES** Chebaling, Nanling (Guangdong), Jianfengling (Hainan), Jiulianshan (Jiangxi). **CONSERVATION** Least Concern. May be threatened by habitat loss.

Grey-capped Pygmy Woodpecker ▪ *Dendrocopos canicapillus*
星头啄木鸟 (Xīng tóu zhuó mù niǎo) 15cm

DESCRIPTION ssp. *kaleensis* (shown), *scintilliceps* (E to C China), *swinhoei* (Hainan). One of the smallest woodpeckers. Upperparts mostly brownish-black with white back and wing spotting, contrasting with pale, finely streaked underparts. Crown to nape dark grey; brownish facial patch merges with malar, giving a white-moustached appearance. Female (shown) lacks red crown spot. **DISTRIBUTION** Himalayan foothills, NE India, across to E China, Russian Far East and Korea. Also Southeast Asia. Widespread SE China. **HABITS AND HABITATS** Common resident of broadleaved evergreen forests, forest edges from hills to 1,900m, and sometimes woodland. Frequently in mixed flocks with other woodpeckers, tits and warblers. Calls a soft *cheep* and a short *trrrrrr*. **TIMING** All year round. **SITES** Nanling (Guangdong), Jiangfengling (Hainan), Jiulianshan, Wuyishan (Jiangxi). **CONSERVATION** Least Concern.

Rufous Woodpecker
▪ *Micropternus brachyurus*
栗啄木鸟 (Lì zhuó mù niǎo) 25 cm

DESCRIPTION ssp. *fokiensis* (shown), *holroydi* (Hainan). Reddish-brown woodpecker with a slight crest; resembles **Pale-headed Woodpecker**. Head pale with fine streaking on crown and throat. Rest of plumage rufous-brown with fine black barring, especially on upperparts and tail. Male (shown) has a red patch below the eye. **DISTRIBUTION** Himalayan foothills, S and E India, to across S China. Also Southeast Asia. Widespread SE China, from Hainan to Zhejiang. **HABITS AND HABITATS** Uncommon resident of broadleaved evergreen forests and forest edges from lowlands to 1,500m. Usually seen in pairs or small groups. Specialist feeder on ants, also nesting in ants' nests (extralimital data). Calls a whinnying trill, more rapid towards the end. **TIMING** All year round. **SITES** Fuzhou Forest Park (Fujian), Nanling (Guangdong), Jianfengling (Hainan), Jiulianshan (Jiangxi). **CONSERVATION** Least Concern.

Bay Woodpecker ■ *Blythipicus pyrrhotis* 黄嘴栗啄木鸟
(Huáng zuǐ lì zhuó mù niǎo) 30cm

DESCRIPTION ssp. *sinensis* (shown), *hainanus* (Hainan). Rich-brown woodpecker with a bright yellow bill. Head pale brown; rest of upperparts, breast and belly rufous-

brown with black barring. Male (shown) has a bright red nape-patch. **DISTRIBUTION** Himalayan foothills, NE India, mainland Southeast Asia, to S China. In SE China, from Hainan to Fujian. **HABITS AND HABITATS** Uncommon resident of broadleaved evergreen forests from hills to about 1,900m. Shy; often seen in pairs or as single birds. Calls a series of *hee* notes, descending and more rapid towards the end. **TIMING** All year round. **SITES** Emeifeng (Fujian), Nanling (Guangdong), Jianfengling (Hainan). **CONSERVATION** Least Concern. Sensitive to forest fragmentation; now rare in Hong Kong.

Greater Yellownape ■ *Picus flavinucha* 大黄冠啄木鸟
(Dà huáng guān zhuó mù niǎo) 34cm

DESCRIPTION ssp. *ricketti*, *styani* (Hainan). Large and colourful woodpecker. Crest and nape yellow; crown reddish. Head grey with blue orbital skin and white malar patch (yellow in females, shown). Back and wings green; underparts brownish-grey. Similar **Lesser Yellownape** lacks red on crown and has mostly greenish underparts (ssp. *pierrei* from Southeast Asia shown). **DISTRIBUTION** Himalayan foothills, E India to Southeast Asia, S and SE China. In SE China, from Hainan to Fujian. **HABITS AND HABITATS** Rare to locally common resident of broadleaved evergreen forests from 800 to 2,000m. Calls a single *kyep* or an accelerating series of *kik* notes. Forms mixed flocks with other woodpeckers and magpies. **TIMING** All year round. **SITES** Longqishan (Fujian), Jiangfengling (Hainan), Wuyishan (Jiangxi). **CONSERVATION** Least Concern.

Pale-headed Woodpecker ■ *Gecinulus grantia* 竹啄木鸟
(Zhú zhuó mù niǎo) 25cm

DESCRIPTION ssp. *viridanus*. Medium-sized, brownish-green woodpecker. Head pale buff with pink crown-patch in male. Underparts and mantle olive-green. Back, wings and tail washed rufous-brown, with dark barring on flight feathers and tail. **DISTRIBUTION** E Himalayan foothills to Southeast Asia, S, SW and SE China. In SE China, Jiangxi to Fujian. **HABITS AND HABITATS** Uncommon resident of broadleaved evergreen and mixed forests, usually in bamboo thickets from 400 to 1,000m; also disturbed forests. Usually seen in pairs, foraging low in dense bamboo clumps. **TIMING** All year round. **SITES** Fuzhou Forest Park (Fujian), Wuyishan and Yangjifeng (Jiangxi). **CONSERVATION** Least Concern.

Fairy Pitta ■ *Pitta nympha* 仙八色鸫 (Xiān bā sè dōng) 19cm

DESCRIPTION Brightly coloured pitta. Crown rich chestnut, buff brow and black patch across eye to nape. Rest of upperparts green, with blue scapulars. Underparts pale buff; lower belly and vent red.
DISTRIBUTION Breeds Korea, S Japan, Taiwan to E, SE China. Winters mostly in Borneo, and possibly mainland Southeast Asia. **HABITS AND HABITATS** Uncommon summer visitor to broadleaved evergreen and mixed forests from lowlands up to 1,400m; sometimes in open areas, including scrub during migration Forages on the ground for soft-bodied invertebrates. Calls a ringing, disyllabic *wee-wiu, wee-wiu*. **TIMING** Late April–September. **SITES** Jiulianshan (Jiangxi), Chebaling, Nanling (Guangdong). **CONSERVATION** Vulnerable. Class II protected species. Threatened by habitat loss in breeding and wintering ranges.

Large Woodshrike ■ *Tephrodornis virgatus* 钩嘴林鵙 (Gōu zuǐ lín jú) 20cm

DESCRIPTION ssp. *latouchei*, *hainanus* (Hainan, shown) Chunky and shrike-like. Crown to nape grey; rest of upperparts greyish-brown, with a white rump. Underparts

white. Black facial mask in male; greyish-brown in female. Female duller brown than male on upperparts, including on crown. **DISTRIBUTION** Himalayan foothills, E and NE India, to Southeast Asia and S China. In SE China, from Hainan to Fujian. **HABITS AND HABITATS** Uncommon resident of broadleaved evergreen forests and forest edges mostly in lowlands and hills, but up to 1,500m. Gregarious; often in small flocks, sometimes joining mixed flocks with monarchs and woodpeckers. Calls a rapid series of *whi* notes; also a harsh *crrrch*. **TIMING** All year round. **SITES** Jianfengling (Hainan), Jiulianshan, Wuyishan (Jiangxi). **CONSERVATION** Least Concern.

brown

Large Cuckooshrike ■ *Coracina macei* 大鹃鵙 (Dà juān jú) 27cm

DESCRIPTION ssp. *rexpineti*, *larvivora* (Hainan). Head and upperparts dark grey; underparts paler grey, fading to white on lower belly and vent. Male has a black facial mask. Female has a black eye-stripe; less strongly marked than male; finely barred on breast

and rump (ssp. *siamensis* from S China shown). **DISTRIBUTION** Indian subcontinent, mainland Southeast Asia to S, SE China. In SE China, from Hainan to Fujian. **HABITS AND HABITATS** Rare to locally uncommon resident in broadleaved evergreen and mixed forests from lowlands up to 1,200m. Sometimes in forest edges and cultivated areas. Forages in canopy, occasionally joining mixed flocks. Calls a raspy, nasal *kwee-it*. **TIMING** All year round. **SITES** Jianfengling (Hainan), Jiulianshan (Jiangxi). **CONSERVATION** Least Concern.

Black-winged Cuckooshrike ■ *Coracina melanoschistos* 暗灰鹃鵙
(Àn huī juān jú) 22cm

DESCRIPTION ssp *intermedia*. Similar to **Large Cuckooshrike**, but smaller and less robust. Plumage mostly dull grey; underparts paler on belly, fading to white on vent. Wings and tail black. Female less strongly marked than male; broken white eye-ring; barred from breast to vent; undertail feathers show conspicuous white tips. **DISTRIBUTION** Himalayan foothills, mainland Southeast Asia to most of E China. Widespread SE China. Northern populations migrate to Southeast Asia and N India. **HABITS AND HABITATS** Fairly common summer breeder and passage migrant to much of region; resident on Hainan. Common in broadleaved evergreen and mixed forests, forest edges and woodland. Forages quietly in canopy, usually

singly or in pairs, sometimes with mixed flocks. Calls a descending *fee-fee-fee-wee*. **TIMING** Mostly late Mar–Oct; all year round in Hainan and S Guangdong. **SITES** Nanling (Guangdong), Jianfengling (Hainan), Chongming Dongtan (Shanghai). **CONSERVATION** Least Concern.

Swinhoe's Minivet ■ *Pericrocotus cantonensis* 小灰山椒鸟
(Xiǎo huī shān jiāo niǎo) 19cm

DESCRIPTION Upperparts greyish-brown; crown to nape black merging with black eye-stripe; forecrown and face white. Underparts washed brown, especially on breast and flanks, rump brownish. Female (shown) duller and less strongly marked than male, with greyish-brown, not black cap. Similar **Ashy Minivet** is larger, grey on upperparts and has white underparts. **DISTRIBUTION** Endemic breeder in C, S and SE China. Winters mainland Southeast Asia. In SE China, from Guangdong to Zhejiang. **HABITS AND HABITATS** Locally common summer breeder in broadleaved evergreen, mixed and coniferous forests from lowlands up to 1,500m. Forages in small parties, regularly joining mixed flocks. Calls a high-pitched twitter. **TIMING** Apr–Oct. **SITES** Dinghushan, Chebaling, (Guangdong), Wuyuan (Jiangxi), Xitianmushan (Zhejiang). **CONSERVATION** Least Concern. May be affected by habitat loss in wintering range.

Grey-chinned Minivet ■ *Pericrocotus solaris* 灰喉山椒鸟
(Huī hóu shān jiāo niǎo) 17–19 cm

DESCRIPTION ssp. *griseogularis*. Brightly coloured minivet. Dark greyish on head and mantle; wings black. Throat pale grey. Underparts and wing-patch orange-red while rump bright red in males. **Scarlet Minivet** is larger, black-headed and deep red on underparts.

Female paler; underparts and wing-patch yellow.
DISTRIBUTION Himalayan foothills, NE India, Southeast Asia to across S China. Widespread in SE China, from Hainan to Fujian. **HABITS AND HABITATS** Common

resident in broadleaved evergreen and mixed forests, and occasionally occurs in coniferous forests from 900 to 1,850m, lower in winter. Forms large flocks in winter, often joining mixed flocks with warblers and tits. Calls a high-pitched *wii-wiit*. **TIMING** All year round. **SITES** Fuzhou Forest Park (Fujian), Nanling (Guangdong), Jianfengling (Hainan). **CONSERVATION** Least Concern.

Tiger Shrike ■ *Lanius tigrinus* 虎纹伯劳 (Hǔ wén bó láo) 19cm

DESCRIPTION Robust-looking shrike. Crown to mantle grey; wings, back and tail chestnut-brown with fine barring; black facial mask. Underparts white with faint barring. Female white browed, and more strongly barred on underparts. Juveniles brown and finely scaled. **DISTRIBUTION** Breeds E China, Russian Far East, Korea and Japan. Winters

Thai-Malay Peninsula and Greater Sundas. **HABITS AND HABITATS** Uncommon passage migrant, occurring in broadleaved evergreen forests, forest edges, farmland and parks. Mostly in lowlands. Prefers more densely vegetated areas than other shrikes. Calls a harsh chattering. **TIMING**

Mostly spring passage Apr–May; autumn passage Aug–Oct. **SITES** Suitable habitat across SE China (e.g. Hangzhou Botanical Gardens, Zhejiang). **CONSERVATION** Least Concern. Decline has been noted in breeding

juvenile

range.

Bull-headed Shrike ■ *Lanius bucephalus* 牛头伯劳 (Niú tóu bó láo) 19cm

DESCRIPTION ssp. *bucephalus*. Head orangey and nape orange-brown, contrasting strongly with grey back and mantle; wings black. Underparts washed orange-brown. Female less strongly marked than male, and finely barred on underparts. Juveniles mostly orange-brown with barrings on underparts. Similar **Brown Shrike** is less richly coloured. **DISTRIBUTION** Breeds C and NE China, Russian Far East, Korea and Japan. Winters mostly E, SE China, S Japan. **HABITS AND HABITATS**

Uncommon passage migrant and winter visitor, occurring in woodland, forest edges, farmland and occasionally parks. Shy, often perching unobtrusively on a low perch. Like other shrikes, impales prey on thorns. Calls a raspy *chiu*, usually followed by harsh *kip* notes. **TIMING** Sep–Apr. **SITES** Fuzhou Forest Park (Fujian), Hangzhou Botanical Gardens (Zhejiang). **CONSERVATION** Least Concern.

juvenile

Brown Shrike ■ *Lanius cristatus* 红尾伯劳 (Hóng wěi bó láo) 19cm

DESCRIPTION Combination of black mask, white brow and forehead distinguishes this species from **Bull-headed Shrike**. Crown pale grey (ssp. *lucionensis*), brown (ssp. *cristatus*; shown, *confusus*), rich brown (ssp. *superciliosus*). Rest of upperparts brown to greyish-brown; tail rufous-brown. Underparts off-white and washed buff, scaled in young birds and females. **DISTRIBUTION** Breeds C, E Siberia to Russian Far East, N China, Korea and Japan. Winters Indian subcontinent, S, SE China and Southeast Asia. **HABITS AND HABITATS** Common passage migrant and winter visitor, uncommon and local breeder (ssp. *lucionensis*), occurring in forest edges, woodland, farmland, scrub and even parks. Highly territorial, even on migration. Calls a scolding *chak-chak-chak* when alarmed. **TIMING** All year round. **SITES** Suitable habitat across SE China. **CONSERVATION** Least Concern.

Long-tailed Shrike ▪ *Lanius schach* 棕背伯劳 (Zōng bèi bó láo) 21–25cm

DESCRIPTION ssp. *schach*. Distinct shrike with a long tail. Head and upper back grey, with a black facial mask. Mantle, wing-coverts and rump rufous-orange; wings and tail black. Underparts white washed rufous on flanks. Juveniles dull brown and finely scaled.

Dark morphed birds mostly grey with black hood and wings.

DISTRIBUTION S Central Asia, Indian subcontinent to S China, Southeast Asia and New Guinea. Widespread SE China. **HABITS AND HABITATS** Common resident of forest edges, farmland, grassland and scrub. Hunts from open perches for insects, small birds and lizards. Calls a variety of harsh screeches and typical, harsh chatters when alarmed. **TIMING** All year round. **SITES** Suitable habitat across SE China. **CONSERVATION** Least Concern.

dark morph

Green Shrike Babbler ▪ *Pteruthius xanthochlorus* 淡绿鵙鹛 (Dàn lǜ jú méi) 13cm

DESCRIPTION ssp. *obscurus*. Together with other shrike babblers, recently reclassified with the New World Vireos. Small, compact and chunky looking bird with a stubby bill.

Head and neck bluish-grey; back and scapulars olive-green. Underparts white, washed yellow at flanks. Both sexes show white eye-ring, but females are less strongly marked. **DISTRIBUTION** Himalayas, N Southeast Asia to SW, S and SE China. In SE China, only in Zhejiang, E Jiangxi and Fujian. **HABITS AND HABITATS** Uncommon resident of broadleaved evergreen forests from 900 to 1,850m. Descends to lower elevations in winter. Commonly joins mixed flocks in pairs or small groups. Calls a quick series of *chi* notes, somewhat tit-like. **TIMING** All year round. **SITES** Wuyishan (Jiangxi), Shenlong Valley (Zhejiang). **CONSERVATION** Least Concern.

Black-naped Oriole ■ *Oriolus chinensis* 黑枕黄鹂 (Hēi zhěn huáng lí) 26cm

DESCRIPTION ssp. *diffusus*. Only bright yellow oriole in region. Plumage rich yellow; thick black eye-band from lore to nape; bright pink bill. Flight feathers mostly black, edged yellow. Female duller than male, and washed dirty yellow on wing-coverts and back. Juveniles heavily streaked. **DISTRIBUTION** Breeds Southeast Asia, S, E, SE China to NE China, Russian Far East and Korea. East Asian populations winter across Southeast Asia and E, NE India **HABITS AND HABITATS** Uncommon summer breeder and passage migrant, occurring in forest edges, deciduous woodland, tree plantations and occasionally parks, mostly in lowlands. Feeds mostly on fruit, but also known to rob nests. Calls variable, but usually a series of fluty whistles. **TIMING** All year round; commoner during summer months. **SITES** Suitable habitat across SE China. **CONSERVATION** Least Concern.

Silver Oriole ■ *Oriolus mellianus* 鹊色鹂 (Què sè lí) 26cm

DESCRIPTION Striking black-and-white oriole with a pale grey bill and white irises. Male (shown) hood and wings glossy black; rest of plumage silvery-white with fine maroon streaks; tail and vent maroon-brown. Female finely streaked on throat and underparts; brownish-black on hood and wings. **DISTRIBUTION** Endemic breeder from C to SE China (Guangdong). Winters in C, SE Thailand and Cambodia. **HABITS AND HABITATS** Rare to uncommon summer breeder in broadleaved evergreen forests from 600 to 1,700m, occasionally in tree plantations. Occurs in pairs or small parties. Forages in canopy, sometimes joining mixed flocks. Calls a fluty whistle and various raspy notes. **TIMING** May–Oct. **SITE** Nanling (Guangdong). **CONSERVATION** Endangered. Threatened by habitat loss in breeding and wintering ranges.

Crow-billed Drongo

■ *Dicrurus annectans* 鸦嘴卷尾
(Yā zuǐ juán wěi) 28cm

DESCRIPTION Large, glossy drongo with a thick bill. Plumage glossy black, with diagnostic shallow-forked tail slightly upcurled at tip. Absence of crest over forecrown separates it from similar **Hair-crested Drongo** or recently moulted **Greater Racket-tailed Drongo** (Hainan). Young birds white spotted from breast to undertail; tail not as prominently upcurled as in adults. **DISTRIBUTION** Breeds NE India, N Southeast Asia to S, SE China. In SE China, only Guangdong, S Hunan and Hainan. Northern populations winter south to Southeast Asia. **HABITS AND HABITATS** Uncommon summer breeder, occurring in broadleaved evergreen forests from hills to 1,400m. Occasionally joins mixed species flocks. **TIMING** Mostly April–September, all year round in Hainan. **SITES** Nanling (Guangdong), Jiangfengling (Hainan), Mangshan (Hunan). **CONSERVATION** Least Concern. Likely affected by habitat loss in wintering range.

Black-naped Monarch ■ *Hypothymis azurea* 黑枕王鹟
(Hēi zhěn wáng wēng) 16cm

DESCRIPTION ssp. *styani*. Chunky, large-headed blue bird. Head, back and upper breast dull blue; rest of underparts white. Note black tufts on hindcrown and on bill base, and black necklace across breast. Female duller blue than male and lacks the black patches.

DISTRIBUTION Indian subcontinent, S China (including Taiwan) and across Southeast Asia. In SE China, from Hainan to S Fujian. **HABITS AND HABITATS** Common resident of broadleaved evergreen and mixed forests, and forest edges, including secondary woodland from lowlands

to about 900m. Forages by catching insects from perches; also joins mixed flocks. Calls a ringing series of fluid *wee* notes. **TIMING** All year round. **SITES** Sun Yat-sen University (Guangdong), Jianfengling (Hainan), Jiulianshan (Jiangxi). **CONSERVATION** Least Concern.

Red-billed Blue Magpie ■ *Urocissa erythrorhyncha* 红嘴蓝鹊
(Hóng zuǐ lán què) 66cm

DESCRIPTION ssp. *erythrorhyncha*. Large and colourful magpie with a blue mantle, wings and tail. Tail very long with white-tipped feathers. Head to upper breast black, contrasting with silvery-white crown and nape; belly to vent white. Bill and feet red. **DISTRIBUTION** Himalayan foothills to mainland Southeast Asia, C, S, E and SE China. **HABITS AND HABITATS** Common resident of broadleaved evergreen and mixed forests, woodland, farmland to even urban parks, from lowlands up to 1,200m. Usually in small, vocal groups. Calls varied, including harsh screeches and whistles. **TIMING** All year round. **SITES** Present in suitable habitat across SE China (e.g. Wuyishan, Wuyuan, Chebaling). **CONSERVATION** Least Concern.

White-winged Magpie ■ *Urocissa whiteheadi* 白翅蓝鹊
(Bái chì lán què) 46cm

DESCRIPTION ssp. *whiteheadi* (Hainan). Large, greyish-black magpie with a bright orange bill. Upperparts and head black; breast washed smoke-grey; belly white. Wings black with two white wing-bars over coverts. Tail black with white terminus. Irises yellow. Young birds less strongly marked than adults. **DISTRIBUTION** N Indochina, S and SE China (Hainan) **HABITS AND HABITATS** Locally common to rare resident of broadleaved evergreen forests and forest edges from hills to about 1,400m. Lives in family groups, foraging mostly in canopy. Calls various harsh raspy notes with a ringing quality. **TIMING** All year round. **SITES** Bawangling, Jianfengling (Hainan). **CONSERVATION** Least Concern. Likely to be threatened by extensive habitat loss.

bottom: juveniles

Ratchet-tailed Treepie ■ *Temnurus temnurus* 塔尾树鹊
(Tǎ wěi shù què) 30cm

DESCRIPTION All-black treepie with unique 'spiky-sided' tail. Long tail shows 'spike-like' feather extensions due to elongated outer barbs of tail feathers. Head large with

thick, slightly decurved black bill. Irises brownish-red. **DISTRIBUTION** Discontinuously mainland Southeast Asia south to Peninsular Thailand; also SE China (Hainan). **HABITS AND HABITATS** Uncommon to locally common resident of broadleaved evergreen forests and forest edges, up to 1,400m. Usually seen singly or in pairs, sometimes joining mixed flocks. Calls varied, including ringing screams, trills and other metallic notes. **TIMING** All year round. **SITES** Bawangling, Jianfengling (Hainan). **CONSERVATION** Least Concern. Likely to be threatened by extensive habitat loss.

Grey Treepie ■ *Dendrocitta formosae* 灰树鹊 (Huī shù què) 40cm

DESCRIPTION ssp. *sinica* (shown), *insulae* (Hainan). Drab, greyish-brown treepie. Hood to breast grey, contrasting with black face and throat; underparts mostly off-white with chestnut undertail-coverts; rump white. Mantle to scapulars chestnut; wings black with white carpal patch. Bill black and slightly curved. **DISTRIBUTION** Himalayas, NE India, N Southeast Asia to C, SW, S and SE China, including Taiwan. Widespread in SE China, from Hainan to Shanghai. **HABITS AND HABITATS** Common resident, occuring in broadleaved evergreen, mixed and deciduous forests, forest edges and sometimes woodland, mostly from the hills to 1,200m. Forages in pairs or noisy family groups. **TIMING** All year round. **SITES** Suitable habitats in SE China. **CONSERVATION** Least Concern.

Collared Crow ■ *Corvus torquatus* 白颈鸦 (Bái jǐng yā) 52cm

DESCRIPTION Unmistakable large, pied crow. Plumage glossy black with a white collar from nape to upper mantle, neck sides and breast. Young birds less strongly marked than adults, and with white parts replaced by grey. **DISTRIBUTION** N Southeast Asia to S, C, SE China, including Kinmen Island.

Widespread SE China. **HABITS AND HABITATS** Uncommon to locally common resident of open country habitats with scattered trees, including farmland (e.g. paddy fields) and freshwater wetlands, mostly in lowlands; seldom in wooded habitats or urban areas. Like most crows, a scavenger; also robs nests. **TIMING** All year round. **SITES** Futian (Guangdong), Dongting Lake (Hunan), Wuyuan (Jiangxi). **CONSERVATION** Near Threatened. Formerly common and widespread; now in decline for reasons unclear, although habitat loss may have played a part.

Japanese Waxwing

■ *Bombycilla japonica* 小太平鸟
(Xiāo tài píng niǎo) 18cm

DESCRIPTION Smaller of two waxwings. Crest and face warm orange-brown, grading to greyish-brown on mantle and scapulars, and uppertail-coverts. Black eye-stripe extends from lores to rear of crest, but ends on base of crest in **Bohemian Waxwing**, tail ends with reddish bar, not yellow. Yellow stripe along centre of belly. **DISTRIBUTION** Breeds Russian Far East to Sakhalin Island. Winters Korea, Japan and N, C, E and SE China. **HABITS AND HABITATS** Gregarious; usually in dense, nervous flocks feeding in trees with berries. Often in company of **Bohemian Waxwing**. Flight strong and rapid. In winter, inhabits a wide range of wooded areas, including city parks with many trees. Uncommon winter visitor to region. **TIMING** Oct–Apr. **SITES** Wuyuan (Jiangxi), Shanghai parks (Shanghai), Hangzhou Botanical Gardens (Zhejiang). **CONSERVATION** Near Threatened. Trapped for bird trade in large numbers, especially in E China.

▪ TITS ▪

Coal Tit ▪ *Periparus ater* 煤山雀 (Méi shān què) 11cm

DESCRIPTION ssp. *kuatunensis*. Small typical tit. Head, crest and throat black, contrasting with pale cheek-patch. Also small pale patch on nape. Upperparts and wings bluish-grey

with two white bars on wing-coverts. Underparts pale with flanks washed buff. Crest length varies between subspecies. **DISTRIBUTION** NW Europe, N Africa eastwards to E Russia, Japan, Korea, N, E and SE China. In SE China, from Fujian to Zhejiang. **HABITS AND HABITATS** Common resident of mixed and coniferous forests from 1,200 to 2,100m. Forages in small groups, but regularly joins other tits and warbler species to form mixed flocks; keeps to canopy. **TIMING** All year round. **SITES** Wuyishan (Jiangxi), Badagongshan (Hunan), Fengyangshan (Zhejiang). **CONSERVATION** Least Concern.

Yellow-bellied Tit ▪ *Periparus venustulus* 黄腹山雀
(Huáng fù shān què) 10cm

DESCRIPTION Small yellowish tit with a short tail. Male's head and throat to chest black; cheek and hind-neck patch white. Upperparts, wings and tail brownish-yellow with two pale bars on wings; underparts yellow. Female similar to male, but has a white throat and is less strongly marked. **DISTRIBUTION** Widespread C, E, SE China, dispersing widely in winter. Endemic to China. **HABITS AND HABITATS** Common resident of mixed and

deciduous forests; also forest edges, woodland and parkland during winter. Lives in pairs or small groups in summer. Altitudinal migrants descend to lowlands during non-breeding season. Has irrupted to southern China. Feeds on fruits and small insects. **TIMING**

All year round. **SITES** City parks in Guangzhou (Guangdong), Wuyuan (Jiangxi), Badagongshan (Hunan), Hangzhou Botanical Gardens (Zhejiang). **CONSERVATION** Least Concern.

84

Green-backed Tit ■ *Parus monticolus* 绿背山雀 (Lǜ bèi shān què) 13cm

DESCRIPTION ssp. *yunnanensis*. Medium-sized tit. Head black with white cheek and nape-patch. Resembles **Japanese Tit**, but mantle is yellowish-green, and breast to belly is yellow with a black ventral stripe. Wings bluish-black with two white wing-bars. **DISTRIBUTION** Himalayas to N Southeast Asia, SW, C China and Taiwan. Marginal in SE China (W Hunan) **HABITS AND HABITATS** Common resident of broadleaved evergreen and mixed forests, including bamboo-dominated woodland from 1,000 to 2,000m. Forages in small groups. Regularly joins mixed species flocks with other tits and warblers. **TIMING** All year round. **SITES** Badagongshan, Hupingshan (Hunan). **CONSERVATION** Least Concern.

Japanese Tit ■ *Parus minor* 远东山雀 (Yuǎn dōng shān què) 15cm

DESCRIPTION ssp. *minor* (shown), *commixtus*. Medium-sized, familiar tit throughout region. Black on hood and ventral patch stretches to lower belly, contrasting with large white cheek-patches; small nape-patch. Mantle green, scapulars greyish-green and wings black, with pale edging to feathers. Similar **Cinereous Tit** is bluish-grey on mantle. **DISTRIBUTION** S Russian Far East, Japan, Korea to NE, E, C, SE China (except Hainan). Widespread SE China. **HABITS AND HABITATS** Common resident of broadleaved evergreen, mixed and coniferous forests, woodland, tree plantations, scrub and urban parkland, from lowlands to over 2,000m. Usually seen singly or in pairs; occasionally participates in mixed feeding flocks. Song includes harsh raspy notes, broken with 2–3 high-pitched chirps. **TIMING** All year round. **SITES** Suitable habitat across SE China. **CONSERVATION** Least Concern.

Yellow-cheeked Tit ■ *Parus spilonotus* 黄颊山雀
(Huáng jiá shān què) 14cm

DESCRIPTION ssp. *rex*. Medium-sized tit with a conspicuous crest. Male (shown) has a yellow-edged black crest; face to hind-neck yellow, with black eye-stripe extending to neck sides; large black patch extends from throat to belly, contrasting with greyish flanks. Wings black with white spotting on wing-coverts. Female similar to male but more yellowish-green. **DISTRIBUTION** E Himalayas to S China and N Southeast Asia. In SE China, from Guangdong to Zhejiang. **HABITS AND HABITATS** Fairly common resident, occurring in broadleaved evergreen mixed and deciduous forests from 800 to 2,000m. Regularly joins other tits and warblers to form mixed flocks. **TIMING** All year round. **SITES** Wuyishan (Jiangxi), Nanling (Guangdong), Jiulongshan (Zhejiang). **CONSERVATION** Least Concern.

Yellow-browed Tit ■ *Sylviparus modestus* 黄眉林雀
(Huáng méi lín què) 9cm

DESCRIPTION ssp. *modestus*. Small and nondescript warbler-like tit with an indistinct crest. Pale buff-yellow brow and thin eye-ring. Upperparts, wing and tail uniform olive-yellow; underparts buff-yellow. **DISTRIBUTION** Himalayas to SW, C, S China and N Southeast Asia. In SE China, mostly E Jiangxi and Fujian. **HABITS AND HABITATS** Locally common to uncommon resident of broadleaved evergreen mixed and coniferous forests from 1,300 to 1,900m. Occurs in pairs and small family groups, and regularly joins mixed flocks. Calls include thin *tsip* notes and shrill chirps. **TIMING** All year round. **SITES** Wuyishan, Jinggangshan (Jiangxi). **CONSERVATION** Least Concern.

Sultan Tit ▪ *Melanochlora sultanea* 冕雀 (Miǎn què) 20cm

DESCRIPTION ssp. *seorsa*, *flavocristata* (Hainan; shown). Large and striking, black-and-yellow tit. Plumage mostly glossy black. Breast to vent bright yellow. Long, fluffy yellow crest distinct and present in both sexes. Female like male but browner. **DISTRIBUTION** E Himalayan foothills, NE India, Southeast Asia and S China. In SE China, from Hainan to Fujian. **HABITS AND HABITATS** Uncommon resident of broadleaved evergreen and mixed forests from hills to 1,900m. Forms large mixed flocks with minivets, tits and warblers. Calls variable, including a series of melancholy *piu* notes and various nasal squeaks and warbles. **TIMING** All year round. **SITES** Longqishan (Fujian), Bawangling, Jianfengling (Hainan). **CONSERVATION** Least Concern.

Chinese Penduline Tit ▪ *Remiz consobrinus* 中华攀雀 (Zhōng huá pān què) 11cm

DESCRIPTION Small, tit-like bird with facial pattern recalling a shrike. Breeding male has a grey crown and black mask, contrasting with white cheeks and brow. Upperparts and wing-coverts chestnut; wings and tail darker, with pale edging to feathers. Underparts washed buff. Female and non-breeding males less strongly marked, with browner crowns. **DISTRIBUTION** Breeds NE, E China to S Russian Far East. Winters Korea, S Japan and E, SE China. **HABITS AND HABITATS** Uncommon to locally common winter visitor, occurring in a variety of wetland habitats, including reed beds, freshwater and coastal marshes. Calls a repeated, thin *chew*. **TIMING** Oct–Apr. **SITES** Wuyuanwan Wetland Park (Fujian), Gongping Lake (Guangdong), Nanhui Dongtan (Shanghai). **CONSERVATION** Least Concern.

Collared Finchbill ■ *Spizixos semitorques* 领雀嘴鹎 (Lǐng què zuǐ bēi) 23cm

DESCRIPTION ssp. *semitorques*. Large chunky bulbul with a thick, stumpy yellow bill. Hood black; crown and nape dark grey with white-streaked ear-coverts, separated from rest

of olive-green plumage by white half-collar. **DISTRIBUTION** N Southeast Asia to C, S, SE China. Widespread SE China. **HABITS AND HABITATS** Common resident, occurring in various forest types, forest edge, scrub and woodland, mostly below 1,400m. Gregarious, often forming small, noisy groups. Calls various rich *chrup* notes, including a series of repeated *pwit-chweet*. **TIMING** All year round. **SITES** Fuzhou Forest Park (Fujian), Wuyuan (Jiangxi), Hangzhou Botanical Gardens (Zhejiang). **CONSERVATION** Least Concern.

Brown-breasted Bulbul ■ *Pycnonotus xanthorrhous* 黄臀鹎
(Huáng tún bēi) 20cm

DESCRIPTION ssp. *andersoni*. Medium-sized brownish bulbul. Upperparts, wings and tail greyish-brown. Lores, crown to nape black; ear-coverts brownish. Chin to throat white;

breast washed brown. Vent buff-yellow. **DISTRIBUTION** C, S, SE China to N Southeast Asia. Widespread SE China (except Hainan, W Guangdong). **HABITS AND HABITATS** Common resident, occurring in forest edges, scrub and woodland, mostly above 800m, where it is noisy and conspicuous. May undergo some altitudinal migration. Calls include rich *chirrups* and chatters. **TIMING** All year round. **SITES** Wuyuan (Jiangxi), Nanling, Nankunshan (Guangdong), Qingliangfeng Nature Reserve (Zhejiang). **CONSERVATION** Least Concern.

Light-vented Bulbul ■ *Pycnonotus sinensis* 白头鹎 (Bái tóu bēi) 19cm

DESCRIPTION ssp. *sinensis, hainanus*. Familiar bulbul across region. Upperparts brown with olive-green wings; underparts dirty white, washed brown. Crown black; contrasts with prominent white nape band (black in ssp. *hainanus*). Distinct dark band across face broken by buff ear-patch. **DISTRIBUTION** N Southeast Asia, C, S, SE and E China to Ryukyu Islands. Widespread SE China. **HABITS AND HABITATS** Very common resident and migrant of forest edges, woodland, farmland, plantations and parkland, including greenery in urban areas. Often seen perched on wires and bare trees. Gregarious, gathering in large groups of as many as 30–40 birds,

sometimes mixing with **Red-whiskered Bulbuls**. Calls varied, including a ringing *wiit* and a series of 5–6 sweet *chweet* warbles. **TIMING** All year round. **SITES** Present in suitable habitat across SE China. **CONSERVATION** Least Concern.

hainanus

sinensis

Brown-eared Bulbul ■ *Microscelis amaurotis* 栗耳短脚鹎
(Lì ěr duān jiǎo bēi) 20cm

DESCRIPTION ssp. *amaurotis*. Large greyish bulbul with distinct long bill and loose, shaggy crest. Upperparts mostly pale greyish-brown; wings and tail richer brown. Underparts pale grey, washed brown on breast and flanks, and finely streaked white. Large chestnut ear-patch from eye diagnostic. **DISTRIBUTION** Breeds islands off Taiwan and N Philippines, Ryukyu Islands, Korea, Japan. Some northern populations winter south to E and SE China (Zhejiang, Shanghai). **HABITS AND HABITATS** Uncommon winter visitor to forest edges, woodland, scrub, farmland and occasionally urban parks. Gregarious; often in small groups but forms large flocks on migration. Flight undulating. Calls high-pitched and raspy *chreeesp*, and various shrill notes. **TIMING** Nov–Apr. **SITES** Chongming Dongtan (Shanghai), Hangzhou Botanical Gardens (Zhejiang). **CONSERVATION** Least Concern.

Black Bulbul ■ *Hypsipetes leucocephalus* 黑短脚鹎 (Hēi duǎn jiǎo bēi) 25cm

DESCRIPTION ssp. *leucocephalus* (shown), *perniger* (Hainan). Large dark, red-billed bulbul with a loose, shaggy crest and long tail. 'White-headed' form entirely white on hood; rest of plumage black; bill reddish-orange. **DISTRIBUTION** W Himalayas to mainland Southeast Asia, and across C, S, SE China and Taiwan. Widespread SE China. Northern populations migrate south to Southeast Asia, S China, N India. **HABITS AND HABITATS** Locally common resident and altitudinal migrant of broadleaved evergreen and mixed forests from hills to over 2,000m. Highly gregarious; often seen in large flocks. Gathers to feed at fruiting and flowering trees. Calls varied, including a ringing *cheesp* and an upwards inflected *wee-uu*. **TIMING** All year round.

SITES Nanling (Guangdong), Jianfengling (Hainan), Wuyuan (Jiangxi), Hangzhou Botanical Gardens (Zhejiang). **CONSERVATION** Least Concern.

White-headed form

Chestnut Bulbul

■ *Hemixos castonotus* 栗背短脚鹎
(Lì bèi duǎn jiǎo bēi) 21cm

DESCRIPTION ssp. *canipennis*, *castonotus* (Hainan, shown, N Vietnam). Medium-sized, brown and grey bulbul. Forehead, face and neck sides chestnut brown, darker on crown and crest; rest of upperparts chestnut brown while wings and tail dark brown. Throat white and often puffed, rest of underparts washed grey. **DISTRIBUTION** N Vietnam, SE China. In SE China, from Hainan to Fujian. **HABITS AND HABITATS** Common resident, occurring in broadleaved evergreen forest, forest edge and adjacent woodland, from lowlands to 1,600m. Gregarious, often seen in small groups, occasionally joins mixed feeding flocks. Vocal, sings a repeated 'whi-WHI-woo', accompanied by various chatters and churrs. Together with **Mountain Bulbul**, the most frequently seen bulbul in the regions' forests. **TIMING** All year round. **SITES** Fuzhou Forest Park (Fujian), Nanling, Sun Yat-sen University (Guangdong) **CONSERVATION** Least Concern.

Pale Martin ▪ *Riparia diluta* 淡色沙燕 (Dàn sè shā yàn) 12cm

DESCRIPTION ssp. *fohkienensis* (shown), *transbaykalica?*. Small, pale-looking swallow. Upperparts and wings pale greyish-brown. Very similar to **Sand Martin** but breast-band is paler and poorly defined. Also shows less contrast between greyish-brown face and white throat. **DISTRIBUTION** Breeds discontinuously C Asia, C Siberia and Transbaikalia to N Indian subcontinent, C, S, SE China. Winters Indian subcontinent, Southeast Asia, S, SE China. **HABITS AND HABITATS** Common passage migrant, localized and scarce summer breeder, occurring over various freshwater wetland habitats, rivers, riparian woodland and open farmland; breeds mostly in holes dug into sandy lake or river banks. **TIMING** Mostly Apr–Aug. **SITES** Present in suitable habitat across SE China (e.g. Poyang and Dongting Lake). **CONSERVATION** Least Concern.

Asian House Martin ▪ *Delichon dasypus* 烟腹毛脚燕 (Yān fù máo jiǎo yàn) 13cm

DESCRIPTION ssp. *dasypus* (shown), *nigrimentale*. Small swallow with a shallow-forked tail. Upperparts and tail blackish-blue broken by large white rump. Underparts mostly dirty white. Similar **Northern House Martin** is larger, with whiter underparts and a more extensive white rump-patch. **DISTRIBUTION** Breeds Himalayas, Tibetan Plateau, C, S, SE China to Russia Far East, Korea and Japan. Winters S, SE China to across Southeast Asia. **HABITS AND HABITATS** Fairly common passage migrant and winter visitor. Often seen feeding over forests and open country, including agricultural areas, wetlands and sometimes urban areas. Summer breeding birds (ssp. *nigrimentale*) nest in crags in mountains to over 2,000m, or even on man-made structures. **TIMING** All year round, commoner in southern parts of region in winter months. **SITES** Breeds Danxiashan (Guangdong), Wuyishan (Jiangxi); suitable habitats across SE China during non-breeding period. **CONSERVATION** Least Concern.

Asian Stubtail ■ *Urosphena squameiceps* 鳞头树莺 (Lín tóu shù yīng) 10cm

DESCRIPTION Unmistakable tiny, wren-like warbler with a short tail. Upperparts warm brown; underparts off-white, washed brown on flanks and undertail-coverts. Long

buffy-white brow and dark eye-stripe that extends to nape. **DISTRIBUTION** Breeds Russian Far East, Sakhalin Island, Japan, Korea and NE China. Winters S, SE China and mainland Southeast Asia **HABITS AND HABITATS** Uncommon to common winter visitor, occurring in broadleaved and mixed forests; also woodland and scrub during migration. Often skulks in dense areas. Solitary; usually seen foraging quietly on the forest floor or low shrubs. Calls a high-pitched *kyip*, usually when alarmed. **TIMING** Nov–Apr. **SITES** Sun Yat-sen University (Guangdong), Jianfengling (Hainan), Nanhui Dongtan (Shanghai). **CONSERVATION** Least Concern.

Manchurian Bush Warbler ■ *Horornis borealis* 远东树莺 (Yuǎn dōng shù yīng) 15.5cm

DESCRIPTION ssp. *canturians*. Medium-sized, drab brown warbler with a rufous-washed crown. Upperparts rufous-brown with a buffy grey-brown brow, weak eye-stripe and pale face;

underparts pale grey, but whiter on chin and throat, washed buff on breast sides, flanks and undertail-coverts. Similar-looking **Japanese Bush Warbler** is paler on underparts. Tail long, rounded at tip. **DISTRIBUTION** Breeds S Russian Far East, Korea to NE, E China. Winters S, SE China, Taiwan; also N Southeast Asia. **HABITS AND HABITATS** Locally common winter visitor and passage migrant, occurring in woodland, scrub, farmland and well-wooded parks. Difficult to see well, often skulking in dense vegetation. Sings a rich *woo-chi-wi-cheu*, ending abruptly. **TIMING** Nov–Apr. **SITES** Nansha Wetlands (Guangdong), Chongming Dongtan (Shanghai), Xixi National Wetlands Park (Zhejiang). **CONSERVATION** Least Concern.

Brownish-flanked Bush Warbler ■ *Horornis fortipes* 强脚树莺
(Qiáng jiǎo shù yīng) 12cm

DESCRIPTION ssp. *davidiana*. Rather nondescript, uniformly brown warbler with short wings and a longish, slender tail. Upperparts dusky brown; throat to upper belly off-white, increasingly buffy-brown towards lower belly. Flanks to undertail-coverts washed rich brown. Brow buff-brown, with short dark eye-stripe indistinct behind eye.

DISTRIBUTION Himalayas to SW, C, S, SE China; also Taiwan, N Southeast Asia. Widespread SE China. **HABITS AND HABITATS** Common resident, occurring at forest edges, woodland, scrubby hillsides, farmland and thickets from hills to 1,300m. Descends lower in winter. Sings frequently in spring; song is distinctive, starting with a high-pitched *wheeee* followed by *chiwiyou*. **TIMING** All year round. **SITES** Jiulianshan, Wuyishan (Jiangxi), Nanling (Guangdong). **CONSERVATION** Least Concern.

Yellow-bellied Bush Warbler ■ *Horornis acanthizoides* 黄腹树莺
(Huáng fù shù yīng) 10cm

DESCRIPTION ssp. *acanthizoides*. Smallest bush warbler in region; rather compact looking. Mid-brown on upperparts, wing and tail. Rufous-washed on crown; brow buff-brown with dark eye-stripe indistinct behind eyes. Breast, flanks and undertail-coverts buff-yellow. Resembles **Brownish-flanked Bush Warbler**, but smaller and yellowish on underparts.

DISTRIBUTION E Himalayas to SW, S and SE China; also Taiwan. In SE China, Fujian to Zhejiang, and west to Hunan. **HABITS AND HABITATS** Common resident of forest edge, scrubby clearings and bamboo thickets from 1,000 to 2,000m; descends lower in winter. Active when foraging. Skulks in dense vegetation, but vocal in spring. Distinctive song starts with a very high-pitched *see*, followed by a long, ascending trill. **TIMING** All year round. **SITES** Emeifeng (Fujian), Wuyishan, Jinggangshan (Jiangxi). **CONSERVATION** Least Concern.

Rufous-faced Warbler ■ *Abroscopus albogularis* 棕脸鹟莺
(Zōng liǎn wēng yīng) 8cm

DESCRIPTION ssp. *fulvifacies*. Tiny, olive-green warbler with distinct rufous-brown face. Crown olive-green, bounded by two lateral black stripes, thin but broadening towards nape. Rest of upperparts, wings and tail olive-green; underparts generally whitish. Throat

flecked black; thin yellow band across breast. **DISTRIBUTION** NE India, N Southeast Asia to SW, S and SE China; also Taiwan. Widespread SE China. **HABITS AND HABITATS** Common resident of bamboo thickets and scrub in broadleaved and mixed evergreen forests, from 200 to 1,200m, descending lower in winter. Active while foraging; often joins mixed species flocks in winter. **TIMING** All year round. **SITES** Fuzhou Forest Park (Fujian), Wuyuan (Jiangxi), Hangzhou parks (Zhejiang). **CONSERVATION** Least Concern.

Mountain Tailorbird ■ *Phyllergates cucullatus* 金头缝叶莺
(Jīn tóu fèng yè yīng) 12 cm

DESCRIPTION ssp. *coronatus*. Previously thought to be a tailorbird *Orthotomus* sp., but phylogenetic studies have revealed it to be a cettiid warbler. Brightly coloured warbler with a long bill. Head greyish-blue; rest of upperparts olive-green. Cap washed orange, with thin white brow. Throat white; breast to vent bright yellow. **DISTRIBUTION** E Himalayan foothills, NE India, S China and Southeast Asia (including parts of Wallacea). In SE China, occurs in Hainan and Guangdong. **HABITS AND HABITATS** Common resident of broadleaved evergreen and mixed forests, forest edges, woodland and shrubby vegetation usually above 800m. Has colonized remnant woodland and forests in Hong Kong in recent years. Usually in pairs; regularly joins understorey

mixed flocks. Calls a series of 5–6 thin whistles, the fourthat the highest pitch. **TIMING** All year round. **SITES** Guangzhou Botanical Gardens, Nanling (Guangdong), Bawangling (Hainan). **CONSERVATION** Least Concern.

juvenile

Black-throated Bushtit ▪ *Aegithalos concinnus* 红头长尾山雀
(Hóng tóu cháng wěi shān què) 12cm

DESCRIPTION ssp. *concinnus*. Striking head pattern: cap rich orange, extending to nape; black facial mask to neck sides; throat white with distinct black throat-patch. Rest of upperparts greyish-brown. Underparts mostly white, washed orange on flanks; also orange breast-band. Young birds lack black throat-patch. **DISTRIBUTION** Himalayan foothills, NE India to S, C and E China; also Taiwan. Also mainland Southeast Asia. Widespread SE China. **HABITS AND HABITATS** Common resident of broadleaved evergreen mixed and coniferous forests, woodland, scrub and parkland, from lowlands to over 2,000m. Gregarious, forming large flocks of 20 or more birds. Occasionally in mixed flocks with *Phylloscopus* warblers. Calls various shrill, high-pitched notes and trills. **TIMING** All year round. **SITES** Present in suitable habitat across SE China. **CONSERVATION** Least Concern.

Dusky Warbler ▪ *Phylloscopus fuscatus* 褐柳莺(Hè liǔ yīng) 12cm

DESCRIPTION Medium-sized, slim-looking leaf warbler. Greyish-brown on upperparts, wings and tail; throat to breast dirty white, rest of underparts washed buff-brown, undertail-coverts paler. Brow mostly buff, but whitish before eye, contrasting with long black eye-stripe that extends from bill base. **DISTRIBUTION** Breeds C Siberia east to Chukotka, and south to Russian Far East, NE China. Winters S, SE China, NE India and Southeast Asia. **HABITS AND HABITATS** Common winter visitor, occurring in scrub, woodland, farmland and vegetation on fringes of freshwater wetlands (e.g marshes), also mangroves. Very active and vocal. Commonly calls a dry, grating *tak-tak-tak*. **TIMING** Oct–Mar. **SITES** Present in suitable habitat across SE China. **CONSERVATION** Least Concern.

Buff-throated Warbler ■ *Phylloscopus subaffinis* 棕腹柳莺
(Zōng fù liǔ yīng) 11cm

DESCRIPTION Brownish-yellow, somewhat nondescript warbler. Resembles **Dusky Warbler** but brow completely yellowish-buff; upperparts mostly olive-brown, with wings

more strongly olive-green; throat to vent mostly washed yellowish-buff. Sympatric with **Alpine Leaf Warbler**, which is richer yellow, but better separated by song. **DISTRIBUTION** Breeds SW, C & SE China; also N Vietnam. Winters N Southeast Asia, S, SE China. **HABITS AND HABITATS** Common summer breeder, occurring in mixed forests, montane scrub and forest edges from 600 to 1,800m. Wintering birds occur in dense shrubs in woodland, and scrub in hills and lowlands. Song a series of five high-pitched *chi* notes. **TIMING** Breed Apr–Sep; winter Oct–Mar. **SITES** Wuyishan, Jinggangshan (Jiangxi), Hupingshan (Hunan). **CONSERVATION** Least Concern.

Radde's Warbler ■ *Phylloscopus schwarzi* 巨嘴柳莺 (Jù zuǐ liǔ yīng) 13cm

DESCRIPTION Robust-looking warbler with a large-headed appearance. Broad pale brow buffy-brown before and over eyes; bill thick and pinkish-orange. Orange legs appear thick.

Upperparts olive-brown with a greener wash on wings; underparts washed yellow, buffish on flanks; undertail-coverts washed cinnamon-buff. Heavier in build than **Dusky** or **Buff-throated Warbler**. **DISTRIBUTION** Breeds C Siberia to Russian Far East, NE China, Korea. Winters mostly mainland Southeast Asia, SE China. **HABITS AND HABITATS** Uncommon winter visitor and passage migrant, occurring in mixed forests, forest edges, woodland and scrub, often in dense shrubby margins. Calls a *chirri* with a raspy quality. **TIMING** Oct–Mar. **SITES** Chebaling (Guangdong), Dongting Lake (Hunan), Nanhui Dongtan (Shanghai). **CONSERVATION** Least Concern.

Yellow-browed Warbler ■ *Phylloscopus inornatus* 黄眉柳莺
(Huáng méi liǔ yīng) 10.5cm

DESCRIPTION Small, well-marked warbler. Upperparts olive-green; underparts whitish. Brow long and yellowish-buff, grading to white towards nape, contrasting with weak black eye-stripe; rear crown-stripe poorly defined. Tertials pale edged; two wing-bars on greater coverts broader and with dark border. **DISTRIBUTION** Breeds C Siberia to Russian Far East, NE China. Winters NE India, S, SE China and Southeast Asia. **HABITS AND HABITATS** Common winter visitor and passage migrant, occurring in most forest types, woodland, farmland and even urban parks. Very active; often flicks wings and tail. Regularly joins mixed species flocks. Calls a high-pitched *chwewii*, up-slurred at the end. **TIMING** Oct–Mar. **SITES** Present in suitable habitat across SE China. **CONSERVATION** Least Concern.

Arctic Warbler ■ *Phylloscopus borealis* 极北柳莺 (Jí běi liǔ yīng) 12cm

DESCRIPTION ssp. *borealis*. Large, slim-looking leaf warbler. Upperparts dark olive-green; underparts off-white, washed buff-yellow on breast. Head appears long, with long brow to lores, unlike in similar **Two-barred Warbler**. Two thin wing-bars, but one on median coverts often indistinct; tertials not pale on edges. Long primary projection, giving short-tailed appearance. Bill heavy looking and flesh coloured at base. **DISTRIBUTION** Breeds Fennoscandia, NW Russia east to Chukotka, south to Russian Far East, NE China. Winters NE India and much of Southeast Asia. **HABITS AND HABITATS** Common passage migrant in region, occurring in a range of wooded habitats from broadleaved evergreen forests to scrub and parkland, from lowlands to 1,200m. **TIMING** Spring passage Apr–May; autumn passage Sep–Nov. **SITES** Suitable habitat across SE China. **CONSERVATION** Least Concern.

Two-barred Warbler ■ *Phylloscopus plumbeitarsus* 双斑绿柳莺
(Shuāng bān lǜ liǔ yīng) 11.5cm

DESCRIPTION Medium-sized greenish leaf warbler. Long brow extends to meet bill base, contrasting with dark eye-stripe. Crown to upperparts dark olive-green; underparts

washed white. Also two pale wing-bars, but absence of pale edges to tertials separates it from similar **Yellow-browed Leaf Warbler**. Bill two toned, with lower mandible yellow. **DISTRIBUTION** Breeds Transbaikalia, NE Mongolia eastwards to NE China and Korea. Winters S, SE China and Southeast Asia. **HABITS AND HABITATS** Fairly common passage migrant and winter visitor, occurring in most forest types and woodland from low elevations to the hills. Formerly considered a race of **Greenish Warbler**. **TIMING** Oct–Apr. **SITES** Shenzhen (Guangdong), Nanhui Dongtan (Shanghai), Hangzhou Botanical Gardens (Zhejiang). **CONSERVATION** Least Concern.

Eastern Crowned Warbler ■ *Phylloscopus coronatus* 冕柳莺
(Miǎn liǔ yīng) 11.5cm

DESCRIPTION Medium-sized, olive-green leaf warbler with distinct head pattern. Lateral crown-stripe dark, but paler green on median crown-stripe. Upperparts and wings greener

than in **Arctic Warbler**, with a single wing-bar; underparts off-white, washed yellow on undertail-coverts. Bill thick looking, and orange on lower mandible. **DISTRIBUTION** Breeds NE China, Russian Far East, Sakhalin Island, Korea and Japan; also C China. Winters NE India, Southeast Asia. **HABITS AND HABITATS** Uncommon to common passage migrant, occurring in a variety of wooded habitats during migration, including urban parkland. **TIMING** Spring passage Mar–Apr; autumn passage Aug–Oct. **SITES** Sun Yat-sen University (Guangdong), Nanhui Dongtan (Shanghai), Hangzhou Botanical Gardens (Zhejiang). **CONSERVATION** Least Concern.

Hartert's Leaf Warbler ■ *Phylloscopus goodsoni* 华南冠纹柳莺
(Huá nán guān wén liǔ yīng) 10.5cm

DESCRIPTION ssp. *goodsoni* (shown), *fokiensis*. Medium-sized, brightly coloured leaf warbler with two yellowish wing-bars. Resembles **Eastern Crowned Warbler**, but is smaller. Median crown-stripe yellow, broader towards nape. Underparts washed yellow in nominate subspecies, off-white in *fokiensis*. Absence of pale edging on tertials separates it from **Yellow-browed Warbler**. Bill and legs orange. **DISTRIBUTION** Breeds SE China, from Guangdong north to Zhejiang. Winters south to Hainan and S Guangdong. Endemic to China. **HABITS AND HABITATS** Uncommon to locally common summer and winter visitor, occurring in broadleaved evergreen and mixed forests from hills to nearly 1,900m. Often forages by creeping on tree branches, recalling a nuthatch. When breeding, very vocal and has a habit of flicking its wings when singing. Sings a repeated *wit-chee-wiu*. **TIMING** Breed Apr–Oct; winter Nov–Mar. **SITES** Wuyishan (Fujian), Nanling (Guangdong), Jianfengling (Hainan). **CONSERVATION** Least Concern.

Emei Leaf Warbler ■ *Phylloscopus emeiensis* 峨眉柳莺
(É méi liǔ yīng) 11.5cm

DESCRIPTION Medium-sized greenish leaf warbler very similar to **Hartert's Leaf Warbler**, but crown pattern less pronounced. Dark greenish-grey lateral crown-stripe; median crown-stripe paler green, contrasting most strongly towards nape. Rest of upperparts olive-green with flight feathers brighter green; also two yellowish wing-bars. Underparts washed greyish. **DISTRIBUTION** Endemic breeder in C, CE, SE China. Winters probably N Southeast Asia. **HABITS AND HABITATS** Uncommon summer breeder, occurring in broadleaved evergreen and mixed forests from about 1000 to 1,900m. Song is a dry and rapid, high-pitched trill. **TIMING** Apr–Oct. **SITE** Nanling (Guangdong). **CONSERVATION** Least Concern.

Kloss's Leaf Warbler ■ *Phylloscopus ogilviegranti* 白斑尾柳莺
(Bái bān wěi liǔ yīng) 10.5cm

DESCRIPTION ssp. *ogilviegranti*. Medium-sized, greenish-yellow leaf warbler. Similar to **Hartert's Leaf Warbler,** but more compact looking and crown pattern more pronounced.

Brow bright yellow; dark lateral crown-stripe contrasts more strongly with yellow median crown-stripe. Upperparts and wings olive-green, with two pale yellow wing-bars; underparts white, washed yellow on undertail-coverts. **DISTRIBUTION** Breeds N Southeast Asia to C, S, SE China. Winters mainland Southeast Asia south to Cambodia. **HABITS AND HABITATS** Locally common summer breeder, occurring in broadleaved evergreen and mixed forests from 1,000 to 1,850m. Has habit of flicking wings when singing. Sings a high-pitched series of *psit-chi-wit*. **TIMING** Apr–Oct. **SITES** Wuyishan (Fujian), Nanling (Guangdong), Wuyanling (Zhejiang). **CONSERVATION** Least Concern.

Hainan Leaf Warbler ■ *Phylloscopus hainanus* 海南柳莺
(Hái nán liǔ yīng) 11cm

DESCRIPTION Medium-sized yellowish leaf warbler. Unmistakable. Brow and median crown-stripe bright yellow, contrasting with olive lateral crown-stripe. Mantle to wings

olive-green with single yellow wing-bar. Similar to **Hartert's** and **Sulphur-breasted Warbler**, but can be separated by its yellowish ear-coverts. **DISTRIBUTION** Endemic to Hainan Island. **HABITS AND HABITATS** Locally common resident, occurring in broadleaved evergreen forests, including secondary forests, mostly from 600 to 1,500m. Often seen in small groups, regularly joining mixed species flocks. Song a high-pitched series of *psi*, the last note lowest. **TIMING** All year round. **SITES** Bawangling, Jianfengling, Wuzhishan (Hainan). **CONSERVATION** Vulnerable. Threatened by habitat loss due to deforestation.

Sulphur-breasted Warbler ■ *Phylloscopus rickettii* 黑眉柳莺
(Hēi méi liǔ yīng) 10.5cm

DESCRIPTION Medium-sized yellowish leaf-warbler with distinctive crown pattern. Thick greenish-black lateral crown-stripe contrasts with olive-yellow median crown-stripe. Upperparts and wings olive-green with two yellow wing-bars; underparts bright lemon-yellow. Similar **Hartert's Leaf Warbler** has a paler lateral crown-stripe; yellow of underparts less bright. **DISTRIBUTION** Endemic breeder in C and SE China, from Guangdong to Fujian. Winters mainland Southeast Asia south to SE Thailand. **HABITS AND HABITATS** Locally common summer breeder, occurring in broadleaved evergreen and mixed forests from hills up to 1,500m. As passage migrant, may occur in woodland and various open habitats. Regular participant of mixed species flocks. Sings a high-pitched *pee-chee-chiu-chiu*. **TIMING** Apr–Oct. **SITES** Jiulianshan, Wuyishan (Jiangxi), Nanling (Guangdong). **CONSERVATION** Least Concern.

White-spectacled Warbler ■ *Seicercus affinis* 白眶鹟莺
(Bái kuàng wēng yīng) 11.5cm

DESCRIPTION ssp. *intermedius*. Compact-looking, olive-green warbler. Underparts lemon-yellow, richer than in other similar-looking *Seicercus* warblers. Black lateral crown-stripe from forecrown to nape; median crown-stripe and brow grey. Golden-yellow eye-ring broken at the top. Grey on brow replaced by green in some individuals. Rest of upperparts olive-green. **DISTRIBUTION** Discontinuously E Himalayas to SW, C and SE China, N, S Vietnam, N Laos. **HABITS AND HABITATS** Locally common resident, occurring in broadleaved evergreen and mixed forests, often in adjoining bamboo thickets from 1,000 to 1,750m; descends lower in winter. Joins mixed species flocks in non-breeding periods. Song varied and complex; includes a *chi-chwe-chit*, sometimes followed by a trill. **TIMING** Breeds Apr–Oct; winter Nov–Mar. **SITES** Wuyishan (Jiangxi), Nanling (Guangdong). **CONSERVATION** Least Concern.

Bianchi's Warbler ▪ *Seicercus valentini* 比氏鹟莺 (Bǐ shì wēng yīng) 11.5cm

DESCRIPTION Olive-greenish *Seicercus* warbler with a single wing-bar and unbroken eye-ring. Median crown-stripe grey, washed greenish on forehead and lores; lateral crown-stripe black but indistinct on forehead. White

outer-tail feathers more conspicuous than in **Plain-tailed Warbler**. Best distinguished from other *Seicercus* warblers by song. **DISTRIBUTION** Discontinuously C, SW, SE China; also N Vietnam. Winters SW China, N Southeast Asia. **HABITS AND HABITATS** Common summer breeder occurring in broadleaved evergreen and mixed forests, mostly from 1,300 to 1,900m. Active and vocal warbler, foraging in understorey shrubs to canopy. Regular participant of mixed species flocks. Sings a high-pitched *chi-chwe-chit-chwe-chit*, with some variation. **TIMING** Apr–Oct. **SITES** Wuyishan (Jiangxi), Nanling (Guangdong). **CONSERVATION** Least Concern.

Plain-tailed Warbler ▪ *Seicercus soror* 淡尾鹟莺
(Dàn wěi wēng yīng) 11.5cm

DESCRIPTION Olive-green warbler with unbroken eye-ring very similar to **Bianchi's Warbler**, but bill slightly larger, tail shorter and crown pattern less distinct. Yellow

wing-bar often indistinct or absent. Also less white on outer-tail feathers than in **Bianchi's Warbler. DISTRIBUTION** Endemic breeder C, SE China. Winters mainland Southeast Asia. **HABITS AND HABITATS** Common summer breeder, occurring in broadleaved evergreen and mixed forests from low elevations (c. 200m) to 1,400m; altitudinally partitioned from **Bianchi's Warbler** when both species co-occur. Joins mixed species flocks during non-breeding period. Song somewhat similar to Bianchi's Warbler's, but higher pitched. **TIMING** Apr–Oct. **SITES** Wuyishan(Jiangxi), Badagongshan (Hunan). **CONSERVATION** Least Concern.

Chestnut-crowned Warbler ■ *Seicercus castaniceps* 栗头鹟莺
(Lì tóu wēng yīng) 9.5cm

DESCRIPTION ssp. *sinensis*. Unmistakable brightly coloured warbler. Distinct chestnut crown contrasting with grey face and neck sides; black lateral crown-stripe from above eye to nape; throat white. Scapulars olive-green; wings dark green with two yellow wing-bars. Rump and lower underparts yellow. **DISTRIBUTION** E Himalayas, Southeast Asia to Sumatra, S, SW and SE China. Northern populations winter south to S, SE China. **HABITS AND HABITATS** Uncommon summer breeder, occurring in undergrowth and bamboo thickets of broadleaved evergreen and mixed forests (which bamboo should be used?), mostly from 800 to 1,850m, but descending lower in winter. Habit of hovering when foraging for insects. **TIMING** Breeds May–Aug; winters end Oct–Apr. **SITES** Nanling (Guangdong), Badagongshan (Hunan), Jinggangshan, Wuyishan (Jiangxi). **CONSERVATION** Least Concern.

Oriental Reed Warbler
■ *Acrocephalus orientalis* 东方大苇莺
(Dōng fāng dà wěi yīng) 19cm

DESCRIPTION Upperparts greyish-brown, richer on wings; underparts dirty white with brownish wash on flanks and vent. White brow extends beyond eye; note also heavy bill with pink lower mandible. Young birds show some diffused streaking on breast. **DISTRIBUTION** Breeds SC Siberia, Mongolia to Russian Far East, Japan, Korea and E China to as far south as Fujian. Winters E India to S China, Southeast Asia. **HABITS AND HABITATS** Common summer breeder and passage migrant; occurs in wet grassland, marshes, reed beds and farmland with wet areas (e.g. paddy fields); occasionally in mangroves. Generally shy, but breeding birds sing from exposed reed perches. Calls a series of raspy and croaking notes. **TIMING** Mostly Apr–Sep. **SITES** Suitable habitat across SE China (e.g. Xixi Wetlands, Minjiang Estuary). **CONSERVATION** Least Concern.

Blunt-winged Warbler ■ *Acrocephalus concinens* 钝翅苇莺
(Dùn chì wěi yīng) 14cm

DESCRIPTION ssp. *concinens*. Small, plain-looking reed warbler with short wings and a rather long tail. Brow buff, indistinct behind eyes. Upperparts olive-brown with warm

rufous wash on rump and uppertail-coverts. Underparts white washed buff on flanks and belly, grading to orange-brown on undertail-coverts. **DISTRIBUTION** Breeds discontinuously C Asia and NE, E, SE China (Hunan, W Jiangxi). Winters mostly Indian subcontinent, mainland Southeast Asia. **HABITS AND HABITATS** Uncommon summer breeder mostly in western SE China; otherwise mainly a passage migrant, occurring in wet grassland, freshwater marshes and reed beds, including scrub adjoining wetlands. Song varied and rich, containing high-pitched chirps and wheezy notes. **TIMING** Breeds Apr–Jul. **SITES** Hupingshan, Taoyuandong (Hunan) and Wuyishan (Jiangxi). **CONSERVATION** Least Concern.

Brown Bush Warbler ■ *Locustella luteoventris* 棕褐短翅莺
(Zōng hè duǎn chì yīng) 14cm

DESCRIPTION Medium-sized, drab-looking bush warbler with short wings. Pale brow indistinct, sometimes absent. Head to mantle greyish-brown; wings and tail darker

and richer brown. Chin to upper breast off-white; rest of underparts orangey-brown; undertail-coverts unmarked. Tail round ended. Very similar to **Russet Bush Warbler**, but lower mandible is pale. **DISTRIBUTION** E Himalayas, N Southeast Asia to C, SW, S & SE China. **HABITS AND HABITATS** Locally common resident of tall grass, scrub and thickets of forest clearings, mostly above 1,800m; descends lower in winter. Extremely hard to see, skulking in dense vegetation. Best located by distinctive song, which consists of a monotonous series of *tu* notes, sounding somewhat like a steam engine. **TIMING** Breeds Apr–Oct; winters Nov–Mar. **SITES** Hupingshan(Hunan), Wuyishan (Jiangxi). **CONSERVATION** Least Concern.

Russet Bush Warbler ■ *Locustella mandelli* 高山短翅莺
(Gāo shān duǎn chì yīng) 13cm

DESCRIPTION ssp. *melanorhynchus*. Small brownish bush warbler with long, graduated tail. Upperparts dark chestnut-brown. Pale greyish-brown brow extending slightly to behind eye; chin and throat pale with fine streaking. Underparts greyish-white, washed brown on breast and flanks; note also pale crescents on undertail-coverts. **DISTRIBUTION** E Himalayas, C, S, SE China to mainland Southeast Asia. **HABITS AND HABITATS** Uncommon summer breeder and winter visitor, occurring in shrubby edges of forests, tea gardens, bamboo and grassy slopes above 1,000m. Descends in winter to lower hills. Shy and easily overlooked; usually found by unique, insect-like song *cree-ut*. **TIMING** Breeds Mar–Aug; winters Nov–Mar. **SITES** Wuyishan (Jiangxi), Taoyuandong, Badagongshan (Hunan). **CONSERVATION** Least Concern.

Styan's Grasshopper Warbler ■ *Locustella pleskei* 史氏蝗莺
(Shǐ shì huáng yīng) 13.5cm

DESCRIPTION Drab brown warbler with a robust bill. Crown, nape and upperparts greyish-brown with indistinct streaking on mantle. Underparts buff; pale on throat, but warmer on breast and flanks to undertail-coverts. Similar **Middendorff's Warbler** is smaller and warmer brown on wings, with a more slender bill and more distinct brow. **DISTRIBUTION** Breeds small islands off coasts of Russian Far East, NE China, Japan and Korea. Likely to winter coastal SE China and N Vietnam. **HABITS AND HABITATS** Rare, probably overlooked, passage migrant and winter visitor, occurring in freshwater marshes, wet grassland and scrub. Definitive records from region are from Hong Kong's Mai Po Marshes and Shanghai. **TIMING** Nov–Mar. **SITE** Hengsha Island (Shanghai). **CONSERVATION** Vulnerable. Threatened by loss of coastal wetland habitats.

Marsh Grassbird ▪ *Locustella pryeri* 斑背大苇莺 (Bān bèi dà wěi yīng) 13cm

DESCRIPTION ssp. *sinensis*. Medium-sized warbler with a long, graduated tail. Upperparts mid-brown; boldly streaked from crown, mantle and scapulars to uppertail-coverts due to

pale-edged dark feathers. Chin to breast white; rest of underparts washed buff. **DISTRIBUTION** Japan, NE, E and SE China, especially lower Yangtze basin). Some populations winter south to SE China. **HABITS AND HABITATS** Uncommon resident and winter visitor, occurring in freshwater and coastal wetlands, including reed beds and salt marshes, has apparently benefitted from invasive cord grass which now dominates some coastal marshes in region. Forages in dense vegetation. During breeding season sings from exposed reed perches. Song a high-pitched, rapid, trill-like series of *chi* note. **TIMING** All year round. **SITES** Dongting Lake (Hunan), Poyang Lake (Jiangxi), Chongming Dongtan (Shanghai). **CONSERVATION** Near Threatened. Loss of wetland habitats is a major threat.

Striated Prinia ▪ *Prinia crinigera* 山鹪莺 (Shān jiāo yīng) 16cm

DESCRIPTION ssp. *parumstriata*. Large brownish prinia with a long, graduated tail. Upperparts greyish-brown with dark streaking on head, mantle and scapulars. Upperparts

off-white, washed buff on flanks with faint streaking. Wings relatively short. **DISTRIBUTION** Himalayas, NE India, N Southeast Asia, C, S and SE China; also Taiwan. Widespread SE China (except Hainan). **HABITS AND HABITATS** Common resident, occurring in scrub, thickets and grassy clearings in forests from the hills to nearly 1,900m; also scrubby margins of farmland. Sings a rapid, repeated *chiseeseep*. **TIMING** All year round. **SITES** Nanling (Guangdong), Daweishan (Hunan), Wuyuan (Jiangxi). **CONSERVATION** Least Concern.

Hill Prinia ■ *Prinia superciliaris*
黑喉山鹪莺 (Hēi hóu shān jiāo yīng) 18cm

DESCRIPTION ssp. *superciliaris*. Large brownish prinia with a long, thin tail. Head brownish-grey with distinct white brow contrasting with black lore and eye-stripe. Chin to upper breast white, with bold black streaking. Rest of underparts washed buff. **DISTRIBUTION** Southeast Asia to SW, S and SE China. Widespread SE China (except Hainan). **HABITS AND HABITATS** Locally common resident, occurring in forest edges, woodland and shrubby hill slopes, mostly above 600m. Noisy and active, flitting between dense shrubs and constantly twitching tail. Calls include various *churrs*; also a series of frantic *kip* when disturbed. **TIMING** All year round. **SITES** Chebaling, Nankunshan (Guangdong), Jiulianshan (Jiangxi). **CONSERVATION** Least Concern.

Elachura ■ *Elachura formosa* 丽星鹩鹛 (Lì xīng liáo méi) 10cm

DESCRIPTION Tiny, short-tailed songbird; resembles a wren-babbler (hence old name Spotted Wren-babbler), but genetically very distinct and in its own family Elachuridae. Upperparts dark greyish-brown with fine white spotting; wing and tail rufous with broad black bars. Underparts peppery-brown with dense speckles and fine white chevrons. **DISTRIBUTION** E Himalayas, N Southeast Asia to SW, S and SE China. In SE China, patchily from W Hunan to Zhejiang. **HABITS AND HABITATS** Uncommon resident of broadleaved evergreen and mixed forests from hills to 1,800m, often skulking in fern thickets and gullies with dense undergrowth. Song unique, consisting of repeated rising and falling, high-pitched *wiii* notes. **TIMING** All year round, but most common Mar–Jul. **SITES** Emeifeng, Fuzhou Forest Park (Fujian), Wuyishan (Jiangxi). **CONSERVATION** Least Concern.

Pygmy Wren-babbler ■ *Pnoepyga pusilla* 小鳞胸鹪鹛
(Xiǎo lín xiōng jiáo méi) 7.5–9cm

DESCRIPTION ssp. *pusilla*. Recent studies have classified *Pnoepyga* wren babblers into a separate family, Pnoepygidae. Tiny 'wren babbler' with tail-less appearance. Upperparts

greyish-brown with fine buff speckling; wings richer brown with buff spots on coverts. Underparts buffy to tea-yellow with scaling extending to belly. **DISTRIBUTION** C Himalayas, NE India, S, C, SE China eastwards to Lesser Sundas. Widespread SE China, from Hainan to Zhejiang. **HABITS AND HABITATS** Common resident of broadleaved evergreen and mixed forests, mostly above 500m; keeps to forest floor. Sedentary and shy. Song distinctive, consisting of a series of high-pitched, descending *ti— ti—tu*, with well-separated syllables. **TIMING** All year round. **SITES** Nanling (Guangdong), Wuyishan (Jiangxi), Tianmushan (Zhejiang). **CONSERVATION** Least Concern. Has recently colonized Hong Kong's remnant forests.

Streak-breasted Scimitar Babbler ■ *Pomatorhinus ruficollis*
棕颈钩嘴鹛 (Zōng jǐng gōu zuǐ méi) 19cm

DESCRIPTION ssp. *hunanensis, stridulus, nigrostellatus* (Hainan). Medium-sized babbler with a distinct facial pattern. Cap to nape dark brown, contrasting with long white brow; broad black eye-stripe. Slightly decurved yellow bill. Throat and chin white; breast

streaked reddish-brown. **DISTRIBUTION** Himalayas, NE India, Southeast Asia to SW, S, SE China. Widespread SE China. **HABITS AND HABITATS** Common resident in broadleaved evergreen and mixed forests, forest edges, woodland, scrub and bamboo thickets, from hills to 1,900m. Forages near ground

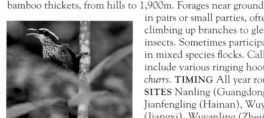

in pairs or small parties, often climbing up branches to glean insects. Sometimes participates in mixed species flocks. Calls include various ringing hoots and *churrs*. **TIMING** All year round. **SITES** Nanling (Guangdong), Jianfengling (Hainan), Wuyishan (Jiangxi), Wuyanling (Zhejiang). **CONSERVATION** Least Concern.

stridulus   *nigrostellatus*

Grey-sided Scimitar Babbler ■ *Pomatorhinus swinhoei*
华南斑胸钩嘴鹛 (Huá nán bān xiōng gōu zuǐ méi) 25 cm

DESCRIPTION spp. *swinhoei* (shown), *abbreviatus*. Large, rufous-brown scimitar babbler with a long, decurved bill. Head and neck dull greyish-brown; rest of upperparts, cheeks and forehead rich chestnut. White lores

and eye-patch. Underparts greyish to dirty white with bold black streaking on breast. **DISTRIBUTION** Widespread SE China from Guangdong to Zhejiang. Endemic to China. **HABITS AND HABITATS** Fairly common resident, occurring in broadleaved evergreen and mixed forests, and frequenting dense undergrowth and scrubby margins of forest edges, from 300 to 1,900m. Skulks in dense vegetation. Calls a series of loud liquid whistles and various *churrs*. **TIMING** All year round. **SITES** Fuzhou Forest Park (Fujian), Nanling (Guangdong), Wuyuan (Jiangxi). **CONSERVATION** Least Concern.

Spot-necked Babbler ■ *Stachyris striolata* 斑颈穗鹛 (Bān jǐng suì méi)16cm

DESCRIPTION ssp. *swinhoei*. Small brown babbler with a distinct facial pattern. White flecking extends from brow to nape and neck sides; facial mask grey with white moustachial patch; throat white. Upperparts chestnut-brown; underparts rich orange-brown. **DISTRIBUTION** Mainland

Southeast Asia to Sumatra; also S and SE China (Hainan) **HABITS AND HABITATS** Common resident of broadleaved evergreen forests and forest edges, including disturbed forests; skulks in dense thickets and vegetated gullies, from lowlands up to c. 1,500m. Often joins mixed flocks of fulvettas and laughingthrushes in forest understorey. Sings a three-note *chee-chi-chee*, the middle note higher. **TIMING** All year round. **SITES** Bawangling, Jianfengling (Hainan). **CONSERVATION** Least Concern. Likely affected by loss of habitat.

Rufous-capped Babbler ■ *Stachyridopsis ruficeps* 红头穗鹛
(Hóng tóu suì méi) 12cm

DESCRIPTION ssp. *davidi* (shown), *goodsoni* (Hainan). Small, olive-brown babbler with an orange-red crown and yellowish throat. Upperparts dark brown; underparts yellow-

buff with fine dark spotting on throat. **DISTRIBUTION** E Himalayas, mainland Southeast Asia to C, S, SE and E China, including Taiwan. Widespread SE China. **HABITS AND HABITATS** Common resident of broadleaved and mixed evergreen forests, and woodland, keeping to dense undergrowth, scrubby fringes and bamboo thickets. Mostly in lowlands and hills, but up to 1,900m. Usually forages relatively low. Regularly joins mixed species flocks. Calls a high-pitched series of *chu* notes, accelerated in the latter syllables. **TIMING** All year round. **SITES** Present in suitable habitat across SE China (e.g. Nanling, Badagongshan). **CONSERVATION** Least Concern.

Dusky Fulvetta ■ *Schoeniparus brunneus* 褐顶雀鹛 (Hè dǐng què méi) 13cm

DESCRIPTION ssp. *superciliaris* (shown), *argutus* (Hainan). Dark, two-toned fulvetta. Crown to most of upperparts brown and finely scaled; face and underparts dull grey; washed brown on flanks and undertail-coverts. Long black brow extends to nape, and

delineates brownish crown from greyish face. **DISTRIBUTION** Widespread C, S & SE China, also Taiwan. Endemic to China. **HABITS AND HABITATS** Common resident, occurring in broadleaved evergreen and mixed forests from 400 to 1,850m, and keeping to shrubby understory. Forages alone or in small groups close to the ground or in leaf litter. Sings a 5 note *fee-fee-wi-chee-wit*, the last three notes faster. **TIMING** All year round. **SITES** Jianfengling (Hainan), Wuyishan (Jiangxi), Taoyuandong (Hunan). **CONSERVATION** Least Concern.

Huet's Fulvetta ■ *Alcippe hueti* 淡眉雀鹛 (Dàn méi què méi) 13cm

DESCRIPTION ssp. *hueti* (shown), *rufescentior* (Hainan). Medium-sized fulvetta with white 'spectacles'. Head to nape grey with black lateral crown-stripe from lores to nape, indistinct in some individuals; large white eye-ring, giving a big-eyed appearance. Rest of upperparts greyish-brown; underparts to undertail-coverts washed buff. Similar to and replaces **David's Fulvetta** in much of region. **DISTRIBUTION** N Southeast Asia, S, SE China. Widespread SE China, from Hainan to Zhejiang. **HABITS AND HABITATS** Common resident of most forest types, forest edges and woodland above 100m; often skulks in dense thickets and forest undergrowth. Regular species in most mixed flocks. Very sensitive to danger, scolding with angry trilling notes when danger (e.g. owl) threatens. Song rich, consisting of 5 notes, *fee-chi-fu-fi-chi*, lowest on final note. **TIMING** All year round. **SITES** Suitable habitat across SE China. **CONSERVATION** Least Concern.

David's Fulvetta ■ *Alcippe davidi* 灰眶雀鹛 (Huī kuàng què méi) 13cm

DESCRIPTION Typical brownish fulvetta with a grey head. Head mostly dull grey with a pale eye-ring. Indistinct black crown-stripe extending from lores to nape. Upperparts dark brown; underparts buff. Very similar to **Huet's Fulvetta**, but replaces it further west (e.g. Hunan, W Jiangxi). **DISTRIBUTION** N Vietnam to C, S & SE China. In SE China, mostly Hunan. **HABITS AND HABITATS** Common resident, occurring in various forest types and forest edges from 100m to nearly 2,000m. Noisy and sociable, usually seen in small family groups. Calls agitated churring notes when disturbed. **TIMING** All year round. **SITES** Badagongshan, Hupingshan (Hunan). **CONSERVATION** Least Concern.

Chinese Grassbird ■ *Graminicola striatus* 大草莺 (Dà cǎo yīng) 17cm

DESCRIPTION ssp. *sinicus* (shown), *striatus*. Large, heavily streaked warbler. Crown warm chestnut, hind-neck white, but darker on mantle and scapulars; heavily streaked throughout

due to pale-edged dark feathers. Rump rusty-brown and unstreaked. Tail graduated. Brow short and white, and ear-coverts buff. Underparts white, washed buff on flanks. **DISTRIBUTION** Mainland Southeast Asia, S and SE China. In SE China, patchily in Guangdong, formerly Hainan. **HABITS AND HABITATS** Rare resident, occurring on grassy hillsides, mostly from 300 to 900m. Mostly silent when not breeding, but vocal in breeding season from March to June. Frequently sings from tops of tall grasses, making short-distance flights over territory. Call is a nasal-sounding *ngeeah*, often in dense vegetation. **TIMING** All year round. **SITES** Recent sightings near Shaoguan (Guangdong); generally under-recorded in region. **CONSERVATION** Near Threatened. Dependent on open scrub, so susceptible to habitat succession.

Chinese Babax ■ *Babax lanceolatus* 矛纹草鹛 (Máo wén cǎo méi) 27cm

DESCRIPTION ssp. *latouchei*. Medium-sized, heavily-streaked laughinghthrush with a long tail. Crown and neck sides to rest of upperparts rich dark-brown, with pale-edged feathers

giving streaked appearance; underparts white with bold streaking from breast to belly. Ear-coverts grey; thick brown malar stripe. **DISTRIBUTION** N Southeast Asia to SW, C, S, SE China. In SE China, from Guangdong to Fujian. **HABITS AND HABITATS** Locally common to uncommon resident of disturbed habitats (e.g. scrubby margins, thickets) and dense undergrowth of broadleaved evergreen and mixed forests, mostly above 1,000m. Forages in undergrowth in small groups. Song consists of a *fee-wu, feee-wu*, the second syllable lower. **TIMING** All year round. **SITES** Tong'an Xiaoping Forest Park (Fujian), Badagongshan (Hunan), Wuyuan, Wuyishan (Jiangxi). **CONSERVATION** Least Concern.

Chinese Hwamei ■ *Garrulax canorus* 画眉 (Huà méi) 24cm

DESCRIPTION ssp. *canorus*. Small brownish laughingthrush with a distinctive facial patch. Bluish-white eye-ring and brow extend to nape; bluish facial skin. Upperparts and wings brown with a richer wash on crown and lores; underparts paler brown; fine dark streaking from head to mantle, and breast. **DISTRIBUTION** N Southeast Asia to S, C, E and SE China. Widespread SE China. **HABITS AND HABITATS** Common resident of broadleaved evergreen and mixed forests, forest edges, shrubby thickets, woodland and scrub, from hills to 1,000m. Skulks in dense vegetation, usually in pairs or small groups. Song a series of loud melodious whistles; also various harsh *churrs*. **TIMING** All year round. **SITES** Suitable habitat across SE China (e.g. Wuyuan, Hangzhou). **CONSERVATION** Least Concern. Heavily trapped for pet-bird trade due to its song.

Rufous-cheeked Laughingthrush ■ *Garrulax castanotis* 栗颊噪鹛 (Lì jiá zào méi) 29cm

DESCRIPTION ssp. *castanotis*. Large greyish laughingthrush with a distinct facial pattern. Facial mask and throat black, contrasting with orange-rufous ear-coverts. Crown to nape, and much of underparts grey; lower belly and undertail coverts washed brown. **DISTRIBUTION** N Vietnam and SE China (Hainan). **HABITS AND HABITATS** Fairly common resident of broadleaved evergreen forests and forest edges; skulks in dense thickets and vegetated gullies, usually above 400m. Mostly in small groups, regularly joining mixed feeding flocks with other laughingthrushes and babblers. Song melodious, consisting of varied rising and falling notes, often broken by outbursts of harsh *churrs*. **TIMING** All year round. **SITES** Bawangling, Jianfengling (Hainan). **CONSERVATION** Least Concern. Likely threatened by habitat loss due to deforestation.

Moustached Laughingthrush ■ *Ianthocincla cineracea* 灰翅噪鹛
(Huī chì zào méi) 22cm

DESCRIPTION ssp. *cineracea*. Small brownish laughingthrush with a distinct facial
pattern. Crown greyish; forehead to nape rich rufous contrasting with white face;

thick black malar extending to neck sides.
Upperparts and wings greyish-brown with
black primary coverts and bluish-grey flight
feathers. Underparts buff-brown, paler towards
throat. **DISTRIBUTION** Discontinuously from
NE India to C, S & SE China. In SE China,
from N Guangdong to Zhejiang. Introduced S
Japan. **HABITS AND HABITATS** Uncommon
to locally common resident of broadleaved
evergreen and mixed forests, bamboo thickets,
plantations and occasionally scrub, from 200
to 1,800m; descends lower in winter. Shy
and easily overlooked. **TIMING** All year
round. **SITES** Fuzhou Forest Park (Fujian),
Hupingshan (Hunan), Wuyishan (Jiangxi),
Daqinggu (Zhejiang). **CONSERVATION** Least
Concern.

Masked Laughingthrush ■ *Garrulax perspicillatus* 黑脸噪鹛
(Hēi liǎn zào méi) 30cm

DESCRIPTION Large greyish laughingthrush with a black mask. Head greyish-brown with a
black mask; rest of upperparts and tail greyish-brown; underparts dull buff. Undertail-coverts

rufous-brown. **DISTRIBUTION**
C, N Vietnam to S, SE, C and
E China. Widespread SE China.
HABITS AND HABITATS Common
resident of well-wooded farmland,
thickets, scrubby hillsides, grassy
areas and even urban parkland,
mostly below 600m. Usually forages
on the ground, sometimes moving
to higher branches. Gregarious and
almost always seen in small groups.
TIMING All year around. **SITE**
Suitable habitat across SE China.
CONSERVATION Least Concern.
Perhaps the most conspicuous
laughingthrush across region.

Greater Necklaced Laughingthrush ■ *Garrulax pectoralis*
黑领噪鹛 (Hēi lǐng zào méi) 31cm

DESCRIPTION ssp. *picticollis* (shown), *semitorquatus* (Hainan). Large laughingthrush with a distinct facial pattern. Black eye-stripe and moustache bounds black-streaked white ear-coverts; broad greyish-black band extends from neck sidesbreast, nearly forming a complete 'necklace'. Upperparts rufous; underparts washed buff. **DISTRIBUTION** C Himalayas, Southeast Asia to S Thailand, east to C, S, SE China. Widespread SE China. **HABITS AND HABITATS** Common resident of broadleaved evergreen and mixed forests, woodland and forest edges, usually above 200m. Gregarious; often in small, noisy groups;

flies in long glides one by one when moving through forest. Often sympatric with **Lesser Necklaced Laughingthrush**. Performs a display where birds hop about bowing and spreading their wings while calling. Most frequently encountered laughingthrush in region's forests. **TIMING** All year round. **SITES** Nanling (Guangdong), Jianfengling (Hainan), Wuyanling (Zhejiang). **CONSERVATION** Least Concern.

Swinhoe's Laughingthrush ■ *Dryonastes monachus* 海南噪鹛
(Hǎi nán zào méi) 23cm

DESCRIPTION Small and richly coloured laughingthrush. Mask to upper breast black, with tuft over lores; cap greyish-blue with some white streaking over forecrown. Nape to mantle, scapulars and most of underparts rich chestnut-brown; wings black. Treated by some authorities as a race of **Black-throated Laughingthrush**. **DISTRIBUTION** Endemic to Hainan Island. **HABITS AND HABITATS** Common resident of broadleaved evergreen forests, forest edges and woodland; also

scrub, skulks in dense thickets and vegetated gullies, from 200 to 1,500m. Often joins mixed flocks with babblers and fulvettas. Song consists of 5–6 rich notes, rising and falling; also outbursts of various raspy *churrs*, especially when alarmed. **TIMING** All year round. **SITES** Bawangling, Jianfengling (Hainan). **CONSERVATION** Not evaluated. Likely to be threatened by extensive habitat loss due to deforestation and hunting.

Blue-crowned Laughingthrush ■ *Dryonastes courtoisi* 靛冠噪鹛
(Diàn guān zào méi) 23cm

DESCRIPTION Small and colourful laughingthrush with a distinct facial pattern. Crown to nape bluish-grey; forehead, facial mask and chin black; throat yellow grading to white

below mask. Mantle to much of wing greyish-brown; flight feathers greyish-blue. Underparts washed buff-grey, with paler undertail-coverts. **DISTRIBUTION** SE China (Jiangxi only). Endemic to China. **HABITS AND HABITATS** Locally common resident, occurring in mixed forests, orchards and woodland, often close to human habitation. Forages in trees and on the ground in small groups. Only laughingthrush to breed in small colonies. Birds disperse widely in winter to sites still unknown. **TIMING** Apr–Jul. **SITE** Wuyuan (Jiangxi). **CONSERVATION** Critically Endangered. Entire range is confined to Jiangxi, where woodland habitat is heavily fragmented by cultivation; also heavily trapped for cage-bird trade.

Buffy Laughingthrush ■ *Dryonastes berthemyi* 棕噪鹛
(Zōng zào méi) 28cm

DESCRIPTION ssp. *berthemyi*. Large, rufous-brown laughingthrush. Combination of two-toned bill, black lore and chin, and blue facial skin distinct. Upperparts, much of head

and upper breast rich brown; underparts silvery-grey; undertail-coverts white. **DISTRIBUTION** Endemic to C, SE China. In SE China, from Hunan to Fujian and Zhejiang. **HABITS AND HABITATS** Uncommon to locally common resident of broadleaved evergreen and mixed forests mostly from 600 to 1,900m, descending lower in winter. Gregarious, often foraging on the ground in small, noisy parties. Frequently calls a rich *wu-weeh*. **TIMING** All year round. **SITES** Wuyishan, Sanqingshan (Jiangxi), Taoyuandong (Hunan). **CONSERVATION** Least Concern.

Red-tailed Laughingthrush ■ *Trochalopteron milnei* 赤尾噪鹛
(Chì wěi zào méi) 26cm

DESCRIPTION ssp. *milnei, sinianum*. Medium-sized, colourful laughingthrush. Crown to nape rich orange-brown, contrasting with black brow, lores and chin; facial skin bluish; ear-coverts greyish-brown (Subspecies *sharpei* from S China is shown here has white ear-coverts). Wings and tail rich crimson. Mantle greyish-brown with bold scaling while underparts mostly bluish-grey.

DISTRIBUTION N Southeast Asia (C Vietnam) to S, SE China. In SE China, mainly W Fujian, N Guangdong and S Hunan. **HABITS AND HABITATS** Uncommon to rare resident of broadleaved evergreen forests, often in bamboo thickets, from 1,000 to 2,000m. Usually in pairs or small groups, foraging in understorey. Calls varied; include drawn, airy whistle repeated regularly. **TIMING** All year round. **SITES** Nanling (Guangdong), Mangshan (Hunan) **CONSERVATION** Least Concern, but increasingly rare in region due to heavy trapping for bird trade.

Red-billed Leiothrix ■ *Leiothrix lutea* 红嘴相思鸟
(Hóng zuǐ xiāng sī niǎo) 15.5cm

DESCRIPTION ssp. *lutea, kwangtungensis (shown)*. Medium-sized colourful 'babbler' with a red bill and slightly notched tail. Crown yellowish-green and throat yellow, contrasting with black malar; pale lores and grey cheeks. Upperparts dark olive-green; flight feathers yellow with red outer fringes. Underparts greyish, washed yellow on ventral region.

DISTRIBUTION Pamirs, W Himalayas to N Southeast Asia, C, S and SE China. Widespread SE China. Introduced S Japan and Hawaii. **HABITS AND HABITATS** Common resident of broadleaved evergreen forests, woodland, tree plantations and montane bamboo scrub, from hills to nearly 2,000m. Usually in vigilant and noisy flocks keeping to dense undergrowth. Regularly joins mixed species flocks. **TIMING** All year round. **SITES** Suitable habitat across SE China. **CONSERVATION** Least Concern. Heavily trapped for cage-bird trade.

Golden-breasted Fulvetta ■ *Lioparus chrysotis* 金胸雀鹛
(Jīn xiōng què méi) 11cm

DESCRIPTION ssp. *swinhoii*. Small and brightly coloured, tit-like fulvetta. Forehead, lores, crown and throat black, contrasting with white median crown-stripe and silvery-white ear-coverts; nape grey. Wings black, with orange-yellow patch on secondaries and primaries. Underparts rich orange-yellow. **DISTRIBUTION** E Himalayas, NE India, N Southeast Asia, SW, C, S, SE China. In SE China, Guangdong to Hunan. **HABITS AND HABITATS** Common resident of broadleaved evergreen and mixed forests; also forest edges above 1,000m. Often skulks in dense bamboo thickets. Regularly participates in mixed feeding flocks. Sings a high-pitched, descending *pi-pi-pew*; also dry twittering. **TIMING** All year round. **SITES** Nanling (Guangdong), Mangshan (Hunan). **CONSERVATION** Least Concern.

Grey-hooded Fulvetta ■ *Fulvetta cinereiceps* 灰头雀鹛
(Huī tóu què méi) 12cm

DESCRIPTION ssp. *guttaticollis*, *fucata*. Small pale fulvetta with a large-headed appearance and pale irises. Head to mantle and breast silvery-grey; indistinct lateral crown-stripe stretches from lores to nape; throat streaking diffused. Mantle to wing-coverts rufous-brown; flight feathers greyish-blue. Flanks to lower underparts washed buff-brown (ssp. *cinereiceps* from SW China shown). **DISTRIBUTION** C, SW, SE China. In SE China, mostly Jiangxi and Fujian. Endemic to China. **HABITS AND HABITATS** Common resident of broadleaved evergreen, mixed and coniferous forests and forest edges. Skulks in dense thickets and bamboo, usually above 1,100m. Often joins mixed flocks with other warblers and babblers. Sings a high-pitched *pee-prrrr*. **TIMING** All year round. **SITES** Wuyishan, Yangjifeng (Jiangxi), Badagongshan (Hunan). **CONSERVATION** Least Concern.

Vinous-throated Parrotbill ■ *Sinosuthora webbiana* 棕头鸦雀
(Zōng tóu yā què) 12cm

DESCRIPTION ssp. *suffusus* (shown), *webbiana*. Small, richly coloured parrotbill with a long tail. Head rich chestnut with a short, stubby grey bill. Mantle to underparts greyish-brown; wings dark reddish-brown. **DISTRIBUTION** N Southeast Asia to C, E, NE China, Russian Far East, Korea. Widespread SE China. **HABITS AND HABITATS**
Common resident of forest and woodland edges, scrub, riparian thickets and reed beds, from lowlands to nearly 2,000m. Often encountered in noisy, active groups of more than ten individuals, foraging in dense vegetation for insects, berries and seeds. Calls include various soft chatters and chirrups. **TIMING** All year round.
SITES Dongting Lake (Hunan), Wuyishan (Jiangxi), Nanhui Dongtan (Shanghai), Xixi National Wetland Park(Zhejiang).
CONSERVATION Least Concern.

Golden Parrotbill ■ *Suthora verreauxi* 金色鸦雀 (Jīn sè yā què) 11.5cm

DESCRIPTION ssp. *pallidus*, *craddocki*. Small, orange-brown parrotbill with a long tail. Head to upperparts orange-brown, brighter on crown; underparts white washed rufous on flanks. On face, ear-coverts washed chestnut, moustache and lores white; throat black.
Wings black, with tertials and secondaries chestnut, tipped black.
(ssp. *verreauxi* from C China shown.)
DISTRIBUTION N Southeast Asia to S, C, SE China; also Taiwan. In SE China, from N Guangdong to Hunan. **HABITS AND HABITATS**
Locally common resident, occurring in bamboo thickets and scrub in broadleaved and mixed evergreen forests, from 1,500 to 2,000m, descending lower during winter. Found in pairs or small groups, often skulking in dense bamboo.
TIMING All year round. **SITES** Nanling (Guangdong), Mangshan (Hunan), Wuyishan (Jiangxi).
CONSERVATION Least Concern.

Short-tailed Parrotbill ■ *Neosuthora davidiana* 短尾鸦雀
(Duǎn wěi yā què) 9.5cm

DESCRIPTION ssp. *davidianus*. Small parrotbill with a richly coloured head. Head to mantle chestnut, throat black and bill pale pink. Rest of upperparts greyish-brown;

uppertail-coverts chestnut. Breast off-white; flanks and rest of underparts washed buff. **DISTRIBUTION** Discontinuously mainland Southeast Asia, S and SE China. In SE China, from N Guangdong to Zhejiang. **HABITS AND HABITATS** Uncommon resident, occurring in tall grass and bamboo thickets in broadleaved and mixed evergreen forests, from low 100m to 1,800m, descending lower during winter. Found in pairs or small groups, often skulking in dense bamboo. **TIMING** All year round. **SITES** Nanling (Guangdong), Wuyishan, Wuyuan (Jiangxi), Tianmushan (Zhejiang). **CONSERVATION** Least Concern. Formerly listed as Vulnerable, but down-listed due to ability to utilize disturbed habitats.

Grey-headed Parrotbill
■ *Psittiparus gularis* 灰头鸦雀 (Huī tóu yā què) 18cm

DESCRIPTION ssp. *fokiensis* (shown), *hainanus* (Hainan). Large parrotbill with a grey head and chunky, orange-yellow bill. Thick black border to crown extends from lores to nape; chin black, contrasting with white throat. Mantle to tail brown; underparts washed buff. **DISTRIBUTION** E Himalayas, NE India, mainland Southeast Asia to SW, C, S, SE China. Widespread SE China (except Guangdong). **HABITS AND HABITATS** Common resident of broadleaved evergreen and mixed forests, forest edges, bamboo thickets and scrub, mostly above 400m. Gathers in large, noisy flocks, keeping mostly to canopy. Not fearful of people. Often calls a nasal *chew*, among various chatters. **TIMING** All year round. **SITES** Jianfengling (Hainan), Wuyishan, Wuyuan (Jiangxi), Hangzhou Botanical Gardens (Zhejiang). **CONSERVATION** Least Concern.

Reed Parrotbill ■ *Paradoxornis heudei* 震旦鸦雀 (Zhèn dàn yā què) 19cm

DESCRIPTION Large, big-headed parrotbill with a robust yellow bill. Head grey with a broad black stripe extending from forehead to nape. Lores black contrasting with white forehead and cheek. Wing-coverts and underparts from belly onwards washed rusty-brown. Tail long and graduated, with white tips to three outermost feathers. **DISTRIBUTION** E, SE China. In SE China, mostly coastal Zhejiang and Shanghai. Recently discovered in Henan and Hubei. Endemic to China. **HABITS AND HABITATS** Locally common resident of coastal and freshwater reed beds; occasionally in adjacent wet habitats. Nest a cup subtended over reeds. Uses robust bill to crack reed stems when foraging. Usually in pairs or small groups. **TIMING** All year round. **SITES** Chongming Dongtan, Nanhui Dongtan (Shanghai). **CONSERVATION** Near Threatened. Loss of reed-bed habitats is a key threat.

Japanese White-eye ■ *Zosterops japonicus* 暗绿绣眼鸟
(Àn lǜ xiù yān niǎo) 11cm

DESCRIPTION ssp. *simplex* (shown), *hainanus* (Hainan). Common white-eye of region. Head and upperparts olive-green. Forehead, throat and vent yellow; flanks and breast dirty white. Similar **Chestnut-flanked White-eye** shows chestnut wash on flanks; less distinct on females. **DISTRIBUTION** Breeds across C, E and S China, South Korea and Japan. Northern populations winter S, SE, E China and mainland Southeast Asia. **HABITS AND HABITATS** Common resident and winter visitor, occurring in broadleaved evergreen, deciduous and mixed forests, woodland and scrub, including urban parkland, from lowlands to 1,850m. Usually in large groups, regularly joining mixed flocks. Calls include various high-pitched, shrill twittering. **TIMING** All year round; most common at year end due to influx of northern migrants. **SITES** Suitable habitat across SE China. **CONSERVATION** Least Concern.

White-collared Yuhina ▪ *Yuhina diademata* 白领凤鹛
(Bái lǐng fèng méi) 17.5cm

DESCRIPTION ssp. *diademata*. Greyish yuhina with a distinct crest. Greyish-brown forehead and crest contrast with white patch from behind eye to hind-crown; black on

lores and chin; ear-coverts streaked brown. Nape to mantle and wing-coverts greyish-brown; wings and tail mostly bluish-grey. **DISTRIBUTION** N Southeast Asia to SW, C, S, SE China. Marginal in SE China, mainly Hunan. **HABITS AND HABITATS** Common resident of broadleaved evergreen and mixed forests, and forest edges, mostly above 1,200m, often in scrubby thickets; descends lower in winter. Forages in pairs or single-species groups. Calls a melancholy *priii*, somewhat similar to a white-eye's. **TIMING** All year round. **SITES** Badagongshan, Hupingshan (Hunan). **CONSERVATION** Least Concern.

Black-chinned Yuhina ▪ *Yuhina nigrimenta* 黑颏凤鹛
(Hēi é fèng méi) 11cm

DESCRIPTION ssp. *pallida*. Small greyish yuhina. Head and crest grey with black lores and chin patch; bill red with black culmen. Upperparts dull greyish-brown; underparts

mostly pale buff with a white throat. **DISTRIBUTION** Himalayas, mainland Southeast Asia to SW, C, S, SE China. Widespread SE China. **HABITS AND HABITATS** Common resident of broadleaved evergreen and mixed forests, usually near scrubby edges, thickets and clearings, mostly above 500m, to nearly 1,850m. Often in small, single-species flocks, occasionally joining other species. Forages mostly in canopy. Sings a rising and falling series of high-pitched *wii* notes. **TIMING** All year round. **SITES** Wuyishan (Jiangxi), Badagongshan (Hunan), Shenlonggu (Zhejiang). **CONSERVATION** Least Concern.

Yellow-billed Nuthatch ▪ *Sitta solangiae* 淡紫䴓 (Dàn zǐ shī) 13cm

DESCRIPTION ssp. *chienfengensis*. Medium-sized blue nuthatch. Head, upperparts and tail violet-blue; underparts grey to dirty white. Forecrown black. Bill, facial skin and irises yellow. Male (shown) has a black post-ocular stripe to nape, absent in female. **DISTRIBUTION** Vietnam, SE Laos and SE China (Hainan). **HABITS AND HABITATS** Uncommon resident, occurring in broadleaved evergreen forests and forest edges, mostly above 900m. Regularly joins mixed species flocks with tits, minivets and warblers, creeping on branches and trunks like other nuthatches. Calls a high-pitched series of rapid *tsit* notes. **TIMING** All year round. **SITES** Jianfengling, Bawangling (Hainan). **CONSERVATION** Near Threatened. Extensive habitat loss throughout its restricted range.

Crested Myna ▪ *Acridotheres cristatellus* 八哥 (Bā gē) 26.5cm

DESCRIPTION ssp. *cristatellus* (shown), *brevipennis* (Hainan). Unmistakable; only myna in region. Plumage entirely black, with a small white wing-patch; shaggy crest over base of bill. Irises red; bill pale ivory. **DISTRIBUTION** C, S, E and SE China, Taiwan, N Southeast Asia. Widely introduced elsewhere (e.g. Philippines, Hawaii). Widespread SE China. **HABITS AND HABITATS** Very common resident of scrub, farmland and urban areas across region. Often in large groups, associating with other starlings (e.g. Black-collared Starling). Calls varied, including diverse shrill and raspy chattering notes. **TIMING** All year round. **SITES** Suitable habitat across SE China. **CONSERVATION** Least Concern. Heavily trapped for cage-bird trade.

White-cheeked Starling ■ *Spodiopsar cineraceus* 丝光椋鸟
(Sī guāng liáng niǎo) 23cm

DESCRIPTION Distinct grey starling with white cheeks and sharp, yellow-orange bill. Hood black; untidy, dirty white crown and cheek-patch. Rest of upperparts greyish-

brown; underparts dirty grey to off-white; white rump and vent. Juveniles are darker. **DISTRIBUTION** Breeds SE Siberia, N and NE China, Russian Far East, Korea and Japan. Winters E, SE China and N Southeast Asia. **HABITS AND HABITATS** Common winter visitor across region, occurring in woodland, scrub, farmland, parkland and mangroves. Has recently bred Hong Kong and Macao. Gregarious; often associates with other starlings in large, noisy flocks. Seldom descends to the ground. Calls varied, including a high-pitched *creee* and raspy chatters. **TIMING** Mostly Sep–Apr. **SITES** Fuzhou Forest Park (Fujian), Futian Reserve (Guangdong), Chongming Dongtan (Shanghai). **CONSERVATION** Least Concern.

Black-collared Starling
■ *Gracupica nigricollis* 黑领椋鸟
(Hēi lǐng liáng niǎo) 28cm

DESCRIPTION Large, well-marked myna with yellow facial skin. Bill black. Head and underparts white, with thick black collar across neck and upper breast. Wings black, with white edging to wing-coverts. **DISTRIBUTION** S and SE China, Taiwan and mainland Southeast Asia. Widespread SE China, from Guangdong to Fujian. **HABITS AND HABITATS** Fairly common to locally very common resident of farmland and urban parks. Associates with **Crested Myna**, where often seen foraging in newly ploughed fields for invertebrates and plant matter. Varied calls include ringing shrieks and harsh chatters, reminiscent of **Crested Myna**. **TIMING** All year round. **SITES** Suitable habitat across SE China (e.g. Wuyuan). **CONSERVATION** Least Concern. Popular bird in cage-bird trade.

White-shouldered Starling

■ *Sturnia sinensis* 灰背椋鸟
(Huī bèi liáng niǎo) 19cm

DESCRIPTION Greyish starling with white shoulder-patches. Irises pale. Head to breast, and back, washed brownish-grey. Rest of underparts white. White shoulders contrast strongly with black wing feathers. Tail black. Female less strongly marked than male, with grey extending to belly. **DISTRIBUTION** Breeds S, SE China, N Vietnam. Widespread SE China, from Guangdong to Zhejiang. Winters S China, Taiwan and mainland Southeast Asia. **HABITS AND HABITATS** Fairly common summer breeder, passage migrant and winter visitor; occurs in scrub, edges of wetlands and farmland, where often seen perched on bare trees and overhead wires. Seldom on the ground. Calls include varied shrill notes and chatters. **TIMING** All year round. **SITES** Xiamen (Fujian), Futian, Gongping Lake (Guangdong), Dongzhaigang (Hainan). **CONSERVATION** Least Concern.

White's Thrush ■ *Zoothera aurea* 怀氏虎鸫 (Huái shì hǔ dōng) 27cm

DESCRIPTION ssp. *aurea*. Unmistakable large thrush with heavily scaled plumage and a robust bill. Crown and nape to most of upperparts brown, covered with bold black scaling due to black-edged brown feathers. Underparts off-white and less boldly scaled, with undertail-coverts whitish. Head appears small relative to body. Dark crescent on ear-coverts, and thin black malars. In flight, note two white wing-patches on underwing. **DISTRIBUTION** Breeds W Russia, C Asia eastwards to Russian Far East, NE China, Japan and Korea. Winters S Japan, S, SE China and Southeast Asia. **HABITS AND HABITATS** Fairly common winter visitor, occurring in various forest types, woodland, well-wooded edges of farmland and urban parkland. Forages mostly on the ground; when flushed, flies to perch in low trees. **TIMING** Sep–Apr. **SITES** Suitable habitat across SE China. **CONSERVATION** Least Concern.

Common Blackbird ▪ *Turdus merula* 乌鸫 (Wū dōng) 29cm

DESCRIPTION ssp. *mandarinus*. Large, very dark thrush with a yellow bill and eye-ring. Adult male sooty-black and lacks the spangling of similar **Blue Whistling Thrush**. Female mostly dark brown with a buff throat. Young birds dark, mostly rufous-brown with pale streaks and mottling on underparts. **DISTRIBUTION** Discontinuously W Europe, Middle East, C Asia eastwards to W, C, E, SE China. Northern populations winter south to S Europe, Middle East, S, SE China and N Southeast Asia. **HABITS AND HABITATS**

Common resident and winter visitor; occurs in various forest types, forest edges, woodland, scrubby margins, thickets and urban parks. Skulks in dense areas, but also feeds in open areas by woodland edges. **TIMING** Most common Nov–Mar due to influx of migrants; smaller numbers during summer months (breeding population). **SITES** Suitable habitat across SE China. **CONSERVATION** Least Concern.

Grey-backed Thrush ▪ *Turdus hortulorum* 灰背鸫 (Huī bèi dōng) 22cm

DESCRIPTION Small greyish thrush. Male mostly grey on head, breast and upperparts; lower breast and flanks rich orange. Female browner on upperparts and spotted black on throat to upper breast. Young birds have spotting extending to orange lower breast. **DISTRIBUTION** Breeds Russian Far East, Korea and NE China. Winters N Vietnam, E and SE China. **HABITS AND HABITATS** Common winter visitor, occurring in

broadleaved evergreen and mixed forests, woodland, shrubby areas and urban parkland. Skulks, keeping to dense vegetation; occasionally forages in open areas during cold weather. **TIMING** Mostly Nov–Mar. **SITE** Suitable habitats in SE China. **CONSERVATION** Least Concern.

Japanese Thrush ■ *Turdus cardis* 乌灰鸫 (Wū huī dōng) 21cm

DESCRIPTION Unmistakable small dark thrush. Male glossy black on head and upperparts, contrasting with white, black-spotted breast and belly; vent white. Bill and eye-ring yellow. First-winter male greyish-blue on back and wings. Female brown on upperparts; underparts off-white; washed buffy-orange on flanks, with chevron-shaped brown spots. **DISTRIBUTION** Breeds Japan, E and C China. Winters SE China, Taiwan, N Southeast Asia south to C Vietnam. **HABITS AND HABITATS** Uncommon to locally common winter visitor to broadleaved evergreen and mixed forests, woodland and sometimes even urban parks, from lowlands to 1,200m. Shy, mostly foraging on forest

floor for arthropods, but also on fruiting trees. **TIMING** Nov–Apr. **SITE** Guangzhou parks (Guangdong), Jianfengling (Hainan), Jinggangshan (Jiangxi), Hangzhou Botanical Gardens (Zhejiang). **CONSERVATION** Least Concern.

Brown-headed Thrush ■ *Turdus chrysolaus* 赤胸鸫 (Chì xiōng dōng) 24cm

DESCRIPTION ssp. *chrysolaus*. Medium-sized, orangey-brown thrush. Adult male mostly mid-brown on upperparts, with a dark brown facial mask. Breast to flanks orange-brown; belly centre to undertail-coverts white. Female similar to male, but has a pale buff brow; throat off-white with streaking. **DISTRIBUTION** Breeds Sakhalin and Kuril Islands, Japan. Winters SE China, Taiwan, Philippines. **HABITS AND HABITATS** Uncommon winter visitor and passage migrant, occurring in broadleaved evergreen and mixed forests, woodland, and occasionally open scrub and farmland. Usually seen gathering to feed at fruiting trees with other thrushes; occasionally descends to forage on the ground. **TIMING** Nov–Mar. **SITES** Xiamen University, Fuzhou Forest Park (Fujian), Nanhui Dongtan (Shanghai). **CONSERVATION** Least Concern.

Pale Thrush ▪ *Turdus pallidus* 白腹鸫 (Bái fù dōng) 23cm

DESCRIPTION Medium-sized, brown-and-grey thrush. Male bluish-grey on hood and tail; upperparts and wings mostly brown with bluish-grey primaries. Underparts washed brown on breast and flanks, but undertail-coverts white. Female similar to male but white on brow, moustache and throat. In flight, note white tips to outer rectrices. **DISTRIBUTION** Breeds Russian Far East, NE China and Korea. Winters Japan, S Korea, E, SE China. **HABITS AND HABITATS** Common winter visitor to various forest types, woodland and even parks and gardens in winter. Forages on the ground for arthropods and fallen

berries, but shy and often skulks in dense herbage. On migration, may form flocks. **TIMING** Nov–Mar. **SITES** Longhushan (Jiangxi), Fuzhou Forest Park (Fujian), Century Park, Nanhui Dongtan (Shanghai). **CONSERVATION** Least Concern.

Blue Whistling Thrush
▪ *Myophonus caeruleus* 紫啸鸫 (Zǐ xiào dōng) 32cm

DESCRIPTION ssp. *caeruleus*. Large dark thrush with a robust black bill. Plumage entirely deep violet, appearing black in poor light. Head to mantle and breast finely streaked with pale spangles. **DISTRIBUTION** C Asia, Himalayas, NE India, Southeast Asia to SW, S, E and SE China. Widespread SE China, from Guangdong to Zhejiang. **HABITS AND HABITATS** Fairly common resident of broadleaved evergreen and mixed forests, forest edges, shrubland and even urban parkland, usually close to rivers and streams. Forages mostly on the ground for worms and arthropods. Calls a long-drawn, shrill whistle. When calling, has habit of fanning tail. **TIMING** All year around. **SITES** Suitable habitats across SE China. **CONSERVATION** Least Concern.

Lesser Shortwing ■ *Brachypteryx leucophris* 白喉短翅鸫
(Bái hóu duān chì dōng) 12cm

DESCRIPTION ssp. *carolinae*. Small chat with a short tail and wings, but long legs. Sexes alike. Plumage mostly olive-brown, paler on belly. Small pale patch over eye extending to lores, absent in similar female **White-browed Shortwing**. **DISTRIBUTION** E Himalayas to Southeast Asia as far as Lesser Sundas. Also S, SW and SE China, from Guangdong to Zhejiang. **HABITS AND HABITATS** Common resident of broadleaved evergreen and mixed forests above 1,000m, often in dense, moist undergrowth. Altitudinal migrant, descends lower in winter. Shy but sings actively during breeding season; song a rich, melodious rising and falling series of high-pitched notes. **TIMING** All year round. **SITES** Jiulianshan (Jiangxi), Tong'an Xiaoping Forest Park (Fujian), Nanling (Guangdong). **CONSERVATION** Least Concern. Recently colonised Hong Kong, probably from populations in Guangdong.

Japanese Robin ■ *Erithacus akahige* 日本歌鸲 (Rì běn gē qú) 15cm

DESCRIPTION ssp. *akahige*. Medium-sized colourful robin. Male's (shown) forehead, and face to upper breast, rich orange; rest of upperparts and wings brown, with a more richly coloured tail; narrow black band across breast, separating from greyish belly. Female resembles male, but lacks black breast-band. **DISTRIBUTION** Breeds Japan, Sakhalin and S Kuril Islands. Winters S, SE China and mainland Southeast Asia. **HABITS AND HABITATS** Uncommon winter visitor, occurring in broadleaved evergreen and mixed forests, forest edges and woodland, and occasionally in parkland during migration, from lowlands to 1,200m. Forages mostly on the ground. Non-vocal in winter. **TIMING** Oct–Apr. **SITES** Fuzhou Forest Park (Fujian), Jianfengling (Hainan), Guangzhou Parks (Guangdong), Xiaoyangshan Island (Zhejiang). **CONSERVATION** Least Concern.

Daurian Redstart ▪ *Phoenicurus auroreus* 北红尾鸲 (Běi hóng wěi qú) 14cm

DESCRIPTION ssp. *auroreus*. Colourful chat of open areas. Male silvery-grey from crown to mantle; hood black; rest of underparts and rump rich orange-rufous. Wings black with large wing-patch. Female mostly brown with smaller white wing-patch; rufous uppertail-coverts and tail.

DISTRIBUTION Breeds SE Siberia to NE China, Russian Far East and Sakhalin Island; also SW, C China. Winters S, E, SE China, Japan, Korea and N Southeast Asia. **HABITS AND HABITATS** Common winter visitor, occurring in open country: shrubland, scrubby hillsides, farmland, wetland edges and urban parks. Perches prominently on exposed positions, returning to same perch after catching prey. **TIMING** Mostly Oct–Mar. **SITES** Suitable habitat across SE China. **CONSERVATION** Least Concern.

Siberian Stonechat ▪ *Saxicola maurus* 黑喉石鵰 (Hēi hóu shí jǐ) 12cm

DESCRIPTION ssp. *stejnegeri*. Most common migratory chat in the region. Breeding male (shown) black hooded; wing and tail blackish; mantle feathers dark brown with pale fringes; throat sides white; breast orange-rufous. Female and non-breeding male less strongly marked and with paler, brownish hood. **DISTRIBUTION** W Russia, C Asia eastwards to E Siberia and NE China; also Himalayas, Tibetan Plateau, SW China. Winters south from Africa, Indian subcontinent to S China and Southeast Asia. E Asian populations may merit full species status, Stejneger's Stonechat (*S. stejnegeri*). **HABITS AND HABITATS** Common passage migrant and winter visitor, occurring in open country: shrubland, scrubby hillsides, farmland and wetland edges. Habit of returning to same perch, which is usually on tops of shrubs, exposed branches or stakes. **TIMING** Late Sep–Apr. **SITES** Suitable habitat across SE China. **CONSERVATION** Least Concern.

Grey Bushchat ■ *Saxicola ferreus* 灰林鵲 (Huī lín jǐ) 14cm

DESCRIPTION ssp. *haringtoni*. Small and dumpy chat. Male (shown) black masked with a white brow and throat; upperparts and breast grey, and belly dirty white. Female mostly brown with a long white brow and white throat. **DISTRIBUTION** W Himalayas, NE India to mainland Southeast Asia, SW, C, S and SE China. Widespread SE China. **HABITS AND HABITATS** Common summer breeder and winter visitor, occurring mostly in open country on hillsides to over 2,000m, especially in scrubby and grassy areas. Descending lower to hills in winter. Often seen on exposed perches, usually low bushes, stakes or rocks. **TIMING** Breeds Apr–Sep; winters Nov–Mar. **SITES** Suitable habitat across SE China. **CONSERVATION** Least Concern.

Orange-flanked Bluetail ■ *Tarsiger cyanurus* 红胁蓝尾鸲
(Hóng xié lán wěi qú) 13.5cm

DESCRIPTION ssp. *cyanurus*. Upperparts entirely blue except for short whitish supercilium. Underparts whitish from throat to vent; flanks washed orange. Female greyish-brown on upperparts with a blue tail; white eye-ring and throat; underparts buff with orange flanks. **DISTRIBUTION** Breeds Fennoscandia, NW Russia eastwards to NE China, Russian Far East and Japan. Winters NE India, S and SE China, Southeast Asia. **HABITS AND HABITATS** Common passage migrant and winter visitor. Occurs in broadleaved evergreen and mixed forests, woodland, scrub and farmland, and occasionally urban parks. Forages in understorey and on the ground for invertebrates. Calls a soft, scratchy *kat*, followed by a thin whistle. **TIMING** Nov–Mar. **SITES** Suitable habitat across SE China. **CONSERVATION** Least Concern.

Rufous-tailed Robin ■ *Luscinia sibilans* 红尾歌鸲 (Hóng wěi gē qú) 13cm

DESCRIPTION Rich-brown robin. Lores and eye-ring buffy-white. Crown to mantle and scapulars greyish-brown; while face, wings and tail rich rufous. Underparts dirty white to

grey with increasingly diffused brown scaling from throat to belly. **DISTRIBUTION** Breeds C Siberia (Krasnoyarsk) eastwards to S Kuril, Sakhalin, Kamchatka. Winters mainland Southeast Asia, S, SE China. **HABITS AND HABITATS** Common winter visitor, occurring in broadleaved evergreen and mixed forests and woodland to about 1,200m; also well-wooded parkland. Forages quietly on forest floor, often in densely vegetated gullies and slopes. Habit of shivering tail. **TIMING** Nov–Apr. **SITES** Nanling, Guangzhou parks (Guangdong), Jianfengling (Hainan). **CONSERVATION** Least Concern.

Oriental Magpie Robin ■ *Copsychus saularis* 鹊鸲 (Què qú) 20cm

DESCRIPTION ssp. *prosthopellus*. Distinctive pied robin with tail often held upright. Head and upperparts glossy black; underparts white. Large white patch across wing. Female duller than male, with black parts replaced by slate-grey. **DISTRIBUTION** Indian subcontinent, Southeast Asia, to C, S and SE China (introduced to Taiwan). Widespread SE China. **HABITS AND HABITATS** Common resident of woodland, scrub, farmland and parkland,

♀

including urban areas. Forages on the ground for invertebrates; highly territorial. Sings a sweet series of rising and falling whistles, usually from an open perch. **TIMING** All year round. **SITES** Suitable habitat across SE China. **CONSERVATION** Least Concern. Commonly trapped for cage-bird trade.

Plumbeous Water Redstart ■ *Rhyacornis fuliginosa* 红尾水鸲
(Hóng wěi shuǐ qú) 12cm

DESCRIPTION ssp. *fuliginosa*. Plumage mostly dark bluish-grey; rump, tail and vent deep chestnut. Female paler than male; mostly grey on upperparts, and dirty white on underparts with grey scaling from chin to belly; rump white. Note also two thin white wing-bars.
DISTRIBUTION Pamirs, W Himalayas, NE India, N Southeast Asia to across C, NE, E and SE China, Taiwan. Widespread SE China. **HABITS AND HABITATS** Common resident, occuring along rivers and streams in

broadleaved evergreen and mixed forests, from hills to about 1,900m. Forages from rocks beside streams and rapids. Frequently fans tail. Song high pitched and rich, with rising and falling notes.
TIMING All year round.
SITES Suitable habitats across China. **CONSERVATION** Least Concern.

Slaty-backed Forktail ■ *Enicurus schistaceus* 灰背燕尾
(Huī bèi yàn wěi) 23cm

DESCRIPTION Crown and nape to mantle slate-grey (appears black in poor light); throat black. White facial patch extends from forehead to beyond eye, unlike in similar
White-crowned Forktail, where white covers only forecrown. Wings black with a broad white patch across coverts; underparts white. **DISTRIBUTION** C Himalayas to mainland South-east Asia, also C, S and SE China. Widespread SE China. **HABITS AND HABITATS** Uncommon resident, occurring along streams and rivers in broadleaved evergreen and mixed forests, from hills to 1,800m. Habit of foraging along roadsides in early mornings; feeds mainly on aquatic insects. Calls a sharp, high-pitched whistle; also metallic rasping notes. **TIMING** All year round.
SITES Emeifeng (Fujian), Jianfengling (Hainan), Jiulianshan, Wuyishan (Jiangxi).
CONSERVATION Least Concern.

White-crowned Forktail ■ *Enicurus leschenaulti* 白冠燕尾
(Bái guān yàn wěi) 27cm

DESCRIPTION ssp. *sinensis*. Largest and darkest forktail. Black hood extends to breast; rest of underparts white. Distinct white crown-patch confined to forehead. Like in other

forktails, wings are black with a broad white patch across the secondary coverts. **DISTRIBUTION** E Himalayas, NE India to C, S and SE China; also Southeast Asia. Widespread SE China. **HABITS AND HABITATS** Fairly common resident, occurring along streams in broadleaved evergreen and mixed forests, from lowlands to 1,200m. Forages for insects along stream sides; also seen feeding at flowering trees. Calls a long, shrill whistle; also a rich, descending series of *chii* notes. **TIMING** All year round. **SITES** Nanling (Guangdong), Wuyuan, Jiulianshan (Jiangxi), Jianfengling (Hainan). **CONSERVATION** Least Concern.

Spotted Forktail ■ *Enicurus maculatus* 斑背燕尾 (Bān bèi yàn wěi) 26cm

DESCRIPTION ssp. *bacatus*. Resembles **White-crowned Forktail** but boldly spotted. Black hood extends to breast; rest of underparts white. Small white forehead-patch. White

spotting from nape to mantle and scapulars. Also broad white wing-patch across secondary coverts. **DISTRIBUTION** W Himalayas to NE India, S and SE China, mainland Southeast Asia. In SE China, from Hunan to Jiangxi, Fujian. **HABITS AND HABITATS** Uncommon resident, occurring along streams in broadleaved evergreen and mixed forests from 500 to 1,900m. Calls a high-pitched, drawn trill with a ringing quality, unlike the whistles of other forktails. **TIMING** All year round. **SITES** Wuyishan (Jiangxi), Nanling (Guangdong), Mangshan (Hunan). **CONSERVATION** Least Concern.

Blue Rock Thrush ■ *Monticola solitarius* 蓝矶鸫 (Lán jī dōng) 22cm

DESCRIPTION ssp. *pandoo*, *philippensis* (shown). Male *pandoo* entirely deep blue with black wings and tail. Male *philippensis* upperparts and hood blue; breast to undertail-coverts rich chestnut. Marked by dense scaling in autumn and winter. Female upperparts dark greyish-brown with faint blue cast; underparts buff with heavy brown scaling. **DISTRIBUTION** Breeds SW Europe, NW Africa eastwards to much of coastal E Asia, Southeast Asia. Some populations winter south to tropics in Africa, Indian subcontinent and Southeast Asia. **HABITS AND HABITATS** Winter visitor (ssp. *philippensis*) and resident (ssp. *pandoo*), occurring in rocky, lightly wooded areas from coasts to over

2,000m; also in settlements. Usually perches atop rocks, trees or rooftops of buildings; site faithful. **TIMING** Most common Sep–Apr due to influx of migrants, but present all year round. **SITES** Yuelushan (Hunan), Qixiling, Wuyishan (Jiangxi), Xiaoyangshan (Zhejiang). **CONSERVATION** Least Concern.

Chestnut-bellied Rock Thrush ■ *Monticola rufiventris* 栗腹矶鸫 (Lì fù jī dōng) 25cm

DESCRIPTION Robust, richly coloured rock thrush. Male deep blue on hood and upperparts; crown shiny blue, contrasting with darker face. Underparts from breast to undertail-coverts rich chestnut, recalling a male niltava. Female dark greyish-brown on upperparts, with strong, pale crescent behind ear-coverts; underparts buff with heavy dark streaking. **DISTRIBUTION** Himalayas, NE India, N Southeast Asia to C, S, SE China. In SE China, mostly Guangdong to Fujian, Jiangxi. Some populations winter N Southeast Asia. **HABITS AND HABITATS** Fairly common summer breeder in mixed and coniferous forests above 1,000m, descending to lower elevations in winter.

Often seen perched on exposed branches and overhead wires. A commonly heard call is a single, harsh *krrrrr*. **TIMING** Breeds Apr–Aug; winters Oct–Mar. **SITES** Emeifeng (Fujian), Nanling (Guangdong), Wuyishan (Jiangxi). **CONSERVATION** Least Concern.

White-throated Rock Thrush ■ *Monticola gularis* 白喉矶鸫
(Bái hóu jī dōng) 18cm

DESCRIPTION Small and colourful rock thrush. Male (shown) cap to nape blue; dark brown across face to most of upperparts; underparts rich rufous-orange. Thin white throat-patch, often inconspicuous. Female brown and heavily scaled; distinguished from female **Chestnut-bellied Rock Thrush**, which is darker and not scaled on back. **DISTRIBUTION** Breeds Russian Far East, NE China and Korea. Winters S, SE China and mainland Southeast Asia. **HABITS AND HABITATS** Uncommon winter visitor and passage migrant; occurs in broadleaved evergreen forests from lowlands to 1,500m, also well wooded parks on migration. Feeds quietly in understorey; often perches still for long periods. Calls a soft, high-pitched *prrsssp*, otherwise mostly silent. **TIMING** Oct–Apr. **SITES** Sun Yat-sen University, Guangzhou (Guangdong), Jianfengling (Hainan). **CONSERVATION** Least Concern. Threatened by habitat loss in wintering grounds.

Brown-chested Jungle Flycatcher ■ *Rhinomyias brunneatus*
白喉林鹟 (Bái hóu lín wēng) 15 cm

DESCRIPTION Large flycatcher with a longish, slightly hooked bill. Upperparts greyish-brown; underparts dirty white with a dark breast-band. Pale loral patch and white throat.

Lower mandible fleshy-orange. Young birds lacks well-demarcated breast-band. **DISTRIBUTION** Endemic breeder in C, S, SE China,from N Guangdong to Zhejiang. Winters Malay Peninsula and Sumatra. **HABITS AND HABITATS** Rare to uncommon summer breeder and passage migrant. Occurs in broadleaved evergreen and mixed forests, from lowlands to nearly 1,800m. Perches low, often picking up prey from the ground. Song has six high-pitched notes; first note highest, descending and finally rising. **TIMING** May–Sep. **SITES** Nanling, Sun Yat-sen University (Guangdong), Mangshan (Hunan), Wuyishan (Jiangxi). **CONSERVATION** Vulnerable. Dependent on forests in wintering grounds, so threatened by habitat loss.

Asian Brown Flycatcher ■ *Muscicapa latirostris* 北灰鹟
(Běi huī wēng) 13cm

DESCRIPTION ssp. *latirostris*. Dull brown flycatcher. Upperparts greyish-brown; underparts dirty white washed brown on breast sides. White eye-ring, pale lores and white moustache, formed by the dark malars. **Dark-sided Flycatcher** appears similar, but is darker, longer winged and has clearer facial patterns.

DISTRIBUTION Breeds across C Siberia to Russian Far East, NE China, Korea and Japan. Disjointed populations in Southeast Asia, Himalayan foothills, C and S India. Winters S India, S China and Southeast Asia. **HABITS AND HABITATS** Fairly common passage migrant and winter visitor, occuring in most forest types, forest edges, woodland and parkland from lowlands to 1,850m. Forages by sallying for insects from a high perch. Calls a rattling series of five descending *chit* notes. **TIMING** Aug–May. **SITES** Suitable sites across SE China. **CONSERVATION** Least Concern.

Ferruginous Flycatcher
■ *Muscicapa ferruginea* 棕尾褐鹟
(Zōng wěi hè wēng) 12.5cm

DESCRIPTION Rich orangey-brown flycatcher. Head bluish-grey; rest of upperparts rich brown; underparts washed rufous-orange. White eye-ring, throat and lores. **DISTRIBUTION** Breeds Himalayan foothills, E India, C, S China and Taiwan, N Southeast Asia. Winters SE China (Hainan), Southeast Asia. **HABITS AND HABITATS** Rare to uncommon passage migrant. Occurs in broadleaved evergreen and mixed forests, woodland and parks. On migration shows up in parks in urban areas. Forages low in understorey. Call a high-pitched *tsiii*. **TIMING** Autumn passage Sep-Nov; spring passage Mar–Apr. **SITES** Sun Yat-sen University, Bawangling (Hainan), Wuyishan (Jiangxi) **CONSERVATION** Least Concern. Dependent on forests in wintering grounds.

Yellow-rumped Flycatcher ■ *Ficedula zanthopygia* 白眉姬鶲
(Bái méi jī wēng) 13cm

DESCRIPTION Bright black-and-yellow flycatcher. Male black on head, back and wings; rump and underparts bright yellow. Also white (not yellow) brow and wing-patch that extends to tertials. Female dull olive-brown on upperparts, with smaller yellow rump-patch than male's; pale underparts faintly washed yellow. **DISTRIBUTION** Breeds Russia Far East, NE China and Korea. Winters Southeast Asia, especially Sumatra and Java. **HABITS**

AND HABITATS Uncommon passage migrant, occuring in woodland, scrub and parkland, including even urban areas on migration. Forages from understorey to canopy, usually near water. Call a hard *trrrrrr* and a weak, high-pitched whistle. **TIMING** Autumn passage Aug–Oct; spring passage Apr. **SITES** Sun Yat-sen University, Nanhui Dongtan (Shanghai). **CONSERVATION** Least Concern.

Narcissus Flycatcher ■ *Ficedula narcissina* 黄眉姬鶲
(Huáng méi jī wēng) 13.5cm

DESCRIPTION ssp. *narcissina* (shown). More richly coloured than **Yellow-rumped Flycatcher**. Male black on upperparts; white wing-patch across coverts; underparts and rump rich orangey-yellow. Brow yellow, not white as in **Yellow-rumped Flycatcher**. Female olive-brown on underparts with a rufous tail; lacks wing patterns. **DISTRIBUTION** Breeds Russian Far East and Japan. Winters Southeast Asia (Borneo, Java). **HABITS**

AND HABITATS Uncommon passage migrant, occurring in forests, woodland, scrub and even parkland on migration. Unobtrusive; forages low in understorey. Calls a soft, thin *pwee* and a bubbly *trrrrrr*. **TIMING** Autumn passage Sep–Nov; spring passage

Mar–Apr. **SITES** South China Botanical Gardens, Sun Yat-sen University (Guangdong), Nanhui Dongtan (Shanghai), Xiaoyangshan (Zhejiang). **CONSERVATION** Least Concern. Dependent on forests in wintering grounds.

Mugimaki Flycatcher ▪ *Ficedula mugimaki* 鸲姬鹟 (Qú jī wēng) 13cm

DESCRIPTION Striking black-and-orange flycatcher. Male black on upperparts with a small white patch over eye; chin to breast washed bright orange. Note also white wing-patch. Female greyish-brown on upperparts, and with two thin wing-bars; throat to upper breast washed pale orange. **DISTRIBUTION** Breeds C Siberia to Russian Far East, NE China and Korea. Winters S China and Southeast Asia. **HABITS AND HABITATS** Uncommon

passage migrant, occurring in forests, woodland, scrub and even parkland while on migration. Forages higher up than other *Ficedula* flycatchers. Calls a harsh, dry chatter, similar to **Yellow-rumped Flycatcher's**. **TIMING** Sep–Apr. **SITES** South China Botanical Gardens (Guangdong), Century Park, Nanhui Dongtan (Shanghai). **CONSERVATION** Least Concern. Dependent on forests in wintering grounds.

Blue-and-white Flycatcher ▪ *Cyanoptila cyanomelana* 白腹姬鹟 (Bái fù lán wēng) 18cm

DESCRIPTION ssp. *cyanomelana* (shown), *intermedia*. Deep blue flycatcher. Males mostly blue on upperparts; face to breast and upper flanks black, rest of underparts white. Recently split from **Zappey's Flycatcher**, which has a deep bluish, not black face. First-winter

males brown with pale throat-patch; blue wings. Female greyish-brown with rich brown back, wings and tail. **DISTRIBUTION** Breeds Russian Far East, NE China, Korea and Japan. Winters Taiwan, Southeast Asia. **HABITS AND HABITATS** Uncommon passage migrant, occurring in forests, woodland, scrub and even parkland

on migration. Forages mainly in canopy. Usually silent, but may call a soft *tick*.

TIMING Spring passage Mar–Apr; autumn passage Sep–Nov. **SITES** Sun Yat-sen University (Guangdong), Nanhui Dongtan (Shanghai). **CONSERVATION** Least Concern. Dependent on forests in wintering grounds.

Hainan Blue Flycatcher ■ *Cyornis hainanus* 海南蓝仙鹟
(Hǎi nán lán xiān wēng) 14cm

DESCRIPTION Striking blue flycatcher with a dark face. Male (shown) upperparts rich blue, with black lores and chin; throat and breast colour varies from blue to bluish-grey; underparts cold greyish. Female greyish-brown from head to mantle; warmer brown

on wings, tail and uppertail-coverts; throat to breast washed orange-brown. Similar **Pale Blue Flycatcher** (Hainan) shows less sharp contrast between blue on breast and grey belly; face paler. **DISTRIBUTION** S, SE China; mainland Southeast Asia south to C Thailand. In SE China, Hainan to Guangdong. Northern populations winter Southeast Asia. **HABITS AND HABITATS** Summer breeder and resident, occuring in broadleaved evergreen and mixed forests, including dense woodland from lowlands to 1,100m. Forages mostly in lower storey. Leaves breeding grounds by September. Song typical of *Cyornis* flycatchers, a rich, varied series of high-pitched notes. **TIMING** Mostly Apr–Sep; all year round in Hainan, S Guangdong. **SITES** Dinghushan, Sun Yat-sen University (Guangdong), Jianfengling (Hainan). **CONSERVATION** Least Concern.

Fujian Niltava ■ *Niltava davidi* 棕腹大仙鹟 (Zōng fù dà xiān wēng) 18cm

DESCRIPTION Large, brilliant blue flycatcher. Male (shown) bluish-black on forehead and mask, with a shiny blue patch on crown and neck sides; rest of upperparts deep blue; flight feathers bluish-grey; underparts orange-rufous. Female greyish-brown on upperparts, warmer on face and wings; tail chestnut; thin shiny blue patch on neck sides. **DISTRIBUTION** Breeds C, S, SE China, N Southeast Asia. Widespread SE China. Winters mainland Southeast Asia south to SE Thailand. **HABITS AND HABITATS** Uncommon to locally common winter visitor, uncommon and local summer breeder of broadleaved evergreen and mixed forests from hills to 1,850m, descends lower in winter, and occasionally showing up in scrub and parks. Seen singly or in pairs. Sings a repeated, high-pitched *piii*. **TIMING** All year round. More records Dec–Apr. **SITES** Nanling, Sun Yat-sen University (Guangdong), Wuyishan (Jiangxi). **CONSERVATION** Least Concern.

Brown Dipper ■ *Cinclus pallasii* 褐河乌 (Hè hé wū) 17cm

DESCRIPTION ssp. *pallasii*. Dark, plump, robin-like bird of fast-flowing waters. Plumage entirely dark brown; tail short. Young birds paler brown than adults (shown), with fine triangular spotting on upperparts and underparts. **DISTRIBUTION** Central Asia (as far west as Afghanistan), Himalayas to N Southeast Asia, C, NE, E, SE China, Taiwan, E Russia and Japan. Widespread in mountains of SE China. **HABITS AND HABITATS** Fairly common resident of fast-flowing streams and rivers in broadleaved and mixed forests, from hills to nearly 1,800m, where it shares habitat with forktails and water redstarts. Perches on rocks or overhanging vegetation. Dives into water to catch insects. Calls a ringing *sweet*. **TIMING** All year round. **SITES** Nanling (Guangdong), Wuyuan, Jiulianshan (Jiangxi). **CONSERVATION** Least Concern.

Orange-bellied Leafbird ■ *Chloropsis hardwickii* 橙腹叶鹎 (Chéng fù yè bēi) 21cm

DESCRIPTION ssp. *melliana* (shown), *lazulina* (Hainan). Only leafbird in region. Male upperparts mostly bright green; breast to vent orangey-yellow. Bluish-black facial mask with blue moustachial stripe. Also blue wings (primaries and primary coverts). Female mostly green with blue moustache. **DISTRIBUTION** Himalayas, mainland Southeast Asia, to across S, SE China. Widespread SE China, from Hainan to Zhejiang. **HABITS AND HABITATS** Common resident of broadleaved evergreen and mixed forests; also forest edges from 400 to 1,800m. Seen singly or in pairs, often at flowering trees. Calls varied, including loud, fluid whistle, *wheet*. Vocal repertoire rich, often mimicking other species. **TIMING** All year round. **SITES** Jianfengling (Hainan), Fuzhou Forest Park (Fujian), Hangzhou Botanical Gardens (Zhejiang). **CONSERVATION** Least Concern. Sometimes trapped for cage-bird trade.

♀

Fire-breasted Flowerpecker ■ *Dicaeum ignipectus* 红胸啄花鸟
(Hóng xiōng zhuó huāniǎo) 9cm

DESCRIPTION ssp. *ignipectus*. Brightly-coloured flowerpecker. Male upperparts glossy blue-black; mask black extending to upper flanks. Underparts washed buff. Bright crimson patch on breast, and black line down dorsum. Female drab yellowish-brown; buff on underparts.

DISTRIBUTION Himalayan foothills, to across C, S, SE China. Also Southeast Asia. Widespread SE China, from Hainan to S Zhejiang. **HABITS AND HABITATS** Common resident of broadleaved evergreen and mixed forests from 800 to 1,800m, sometimes descending to lowlands. Forages for small insects and

♀

mistletoe fruits in canopy; difficult to see well. Song a series of 4 to 5 high-pitched, ringing notes. **TIMING** All year round. **SITES** South China Botanical Gardens (Guangdong), Jianfengling (Hainan), Wuyishan (Jiangxi). **CONSERVATION** Least Concern.

Fork-tailed Sunbird ■ *Aethopyga christinae* 叉尾太阳鸟
(Chā wěi tài yáng niǎo) 9cm

DESCRIPTION ssp. *latouchii* (shown), *christinae* (Hainan). Brightly coloured sunbird with extended tail feathers. Iridescent green crown to nape; black facial mask; chin to upper

breast and neck sides red. Rest of upperparts olive-brown. Female dull olive-brown with faint eye-ring. **DISTRIBUTION** N Southeast Asia (Vietnam), C, S, SE China. Widespread SE China from Hainan to Zhejiang. **HABITS AND HABITATS** Common resident of broadleaved evergreen and mixed forests, forest edges and woodland from lowlands to 1,700m, sometimes entering parkland. Best seen when feeding at flowering trees. Sings a bubbly series of high-pitched *chit* notes. **TIMING** All year round. **SITES** Suitable sites across SE China, including Hangzhou and Guangzhou Botanical Gardens. **CONSERVATION** Least Concern.

Russet Sparrow ■ *Passer rutilans* 山麻雀 (Shān má què) 15cm

DESCRIPTION ssp. *rutilans*. Richly coloured sparrow. Crown, nape, mantle and scapulars rich chestnut (not brown as in **Eurasian Tree Sparrow**). Cheeks dirty white; narrow black throat-patch; rest of underparts grey. Female dull brown with bold streaking on back; pale brow extending to nape. **DISTRIBUTION** Himalayan foothills, N Southeast Asia to across C, E, S, SE China (except Hainan), Korea and Japan. Northern populations winter S Japan, E, SE China. **HABITS AND HABITATS** Common resident and winter visitor, occuring in forest edges, woodland, scrub and farmland, from lowlands to 600m. Wintering birds

often in mixed flocks with **Eurasian Tree Sparrow** in rural areas. Calls include varied chattering notes and chirps, higher pitched than those of other sparrows. **TIMING** All year round. **SITES** Wuyuan (Jiangxi), Gutianshan (Zhejiang). **CONSERVATION** Least Concern.

White-rumped Munia
■ *Lonchura striata* 白腰文鸟 (Bái yāo wén niǎo) 11cm

DESCRIPTION ssp. *swinhoei*. Only munia with a white rump. Upperparts mostly brown; wings and face darker brown. Thin streaking over mantle and scapulars. Throat to breast brown; rest of underparts grey. **DISTRIBUTION** Indian subcontinent, C, S, SE China, Southeast Asia. Widespread SE China. **HABITS AND HABITATS** Common resident of forest edges, woodland, scrub and farmland from lowlands to 1,000m, especially where tall grass is present, where it is sometimes associated with **Scaly-breasted Munia**. Usually seen foraging for grass seeds in small groups. Calls include a bubbly *prrrt* and a high-pitched *ngeee*. **TIMING** All year round. **SITES** Suitable habitat across SE China. **CONSERVATION** Least Concern.

Forest Wagtail ■ *Dendronanthus indicus* 林鹡鸰 (Lín jí líng) 18cm

DESCRIPTION Distinct, strongly marked wagtail. Head to mantle olive-green, with a long white brow. Wings black, with two thick white bars across wing-coverts. Underparts

white, with two thick white patches extending across upper breast. **DISTRIBUTION** Breeds C to NE China, Russian Far East, Korea and Japan. Winters S, SE China, Southeast Asia, NE and S India. **HABITS AND HABITATS** Fairly common passage migrant and winter visitor, occurring in broadleaved evergreen and mixed forests, woodland and tree plantations. Summer breeder in W mountains of Hunan. Habit of swaying body from side to side; when flushed, perches in trees. Call a soft *pink*, which it also makes when in flight. **TIMING** Autumn passage Sep–Oct; spring passage Apr–May. Breeds May–Aug. **SITE** Suitable habitats across SE China. **CONSERVATION** Least Concern. Dependent largely on forests in wintering grounds.

Eastern Yellow Wagtail ■ *Motacilla tschutschensis* 黄鹡鸰 (Huáng jí líng) 18cm

DESCRIPTION ssp. *taivana, macronyx, tschutschensis*. Breeding males olive-green on back; wings black with white edging to feathers; underparts yellow. Crown bluish-grey (ssp. *macronyx*); grey with white brow (ssp. *tschutschensis*); olive-green with yellow brow (ssp.

taivana). Females greyish-brown; throat white, underparts buff. **DISTRIBUTION** Breeds N Mongolia, C Siberia across to Russian Far East, NE China, W Alaska. Winters S, SE China, Taiwan, Southeast Asia (including Wallacea) to New Guinea. **HABITS AND HABITATS** Common winter visitor and passage migrant, occurring on farmland, scrub, wet grassland and edges of wetlands. Often forages on the ground in small groups. Calls a high-pitched di-syllabic *tswee-yip* when flushed. **TIMING** Sep–May. **SITES** Suitable habitat across SE China. **CONSERVATION** Least Concern.

taivana   *tschutschensis*

White Wagtail ■ *Motacilla alba* 白鹡鸰 (Bái jí líng) 19cm

DESCRIPTION ssp. *leucopsis, ocularis* (shown), *baicalensis, lugens*. Distinct pied wagtail; mostly white on underparts, with black nape. *leucopsis*: white face with black throat. *Ocularis*: white face with black eye-stripe, and black chin to throat. *baicalensis*: resembles *leucopsis*, but has a grey mantle. *lugens*: black eye-stripe; extensive black patch on throat and upper breast, which merges with black on upperparts. **DISTRIBUTION** Breeds across Eurasia: W Europe east to E Russia, E China and Korea, south to Himalayas. Winters across tropics and subtropics, including SE China. **HABITS AND HABITATS** Common winter visitor and passage migrant, occurring in woodland, edges of wetlands, scrub, farmland and even urban parks; ssp. *leucopsis* a locally common summer breeder. Usually forages near water. Calls a sharp, ringing *chwee-zit*. **TIMING** Nearly all year round. **SITES** Suitable habitat across SE China. **CONSERVATION** Least Concern.

leucopsis

ocularis

Buff-bellied Pipit ■ *Anthus rubescens* 黄腹鹨 (Huáng fù liù) 15cm

DESCRIPTION ssp. *japonicus*. Small pipit. Lores and chin buff; brow dirty white; thin black malar stretches to form black patch on throat sides. Crown to upperparts greyish-brown with faint streaking; underparts white to buff, with bold black streaking on breast, but thinner towards flanks.

DISTRIBUTION Breeds E Siberia to Russian Far East, Chukotka; also North America. Winters Japan, E, SE China, NE India and N Southeast Asia. **HABITS AND HABITATS** Fairly common winter visitor throughout region (except Hainan), occurring on wet grassland and edges of freshwater wetlands; also in damp areas by woodland and farmland. Calls a soft *tseep*. **TIMING** Oct–Mar. **SITES** Minjiang Estuary (Fujian), Poyang Lake (Jiangxi), Nanhui Dongtan (Shanghai). **CONSERVATION** Least Concern.

Water Pipit ▪ *Anthus spinoletta* 水鹨 (Shuǐ liù) 16cm

DESCRIPTION ssp. *blakistoni*. Medium-sized pipit. Crown to mantle, and wing-coverts, greyish-brown; wings darker brown, with pale edging to secondary coverts and flight feathers.

Underparts creamy-white with brown streaking, mostly on breast; brown streaks on flanks more diffused. **DISTRIBUTION** Breeds discontinuously W Europe, Caucasus eastwards to C China (Tibetan Plateau), Transbaikalia. Winters N Africa, Middle East to NW India, E and SE China. **HABITS AND HABITATS** Uncommon to locally common winter visitor in northern half of region, scarce or absent south (e.g. Hainan), occurring in coastal and freshwater wetlands, edges of lakes and rivers, and wet grassland. Usually forages by edge of water as its name suggests. Calls include a *psee* and *chweep*, of ringing quality. **TIMING** Oct–Mar. **SITES** Yanghu Wetlands (Hunan), Poyang Lake (Jiangxi), Nanhui Dongtan (Shanghai). **CONSERVATION** Least Concern.

Upland Pipit ▪ *Anthus sylvanus* 山鹨 (Shān liù) 17cm

DESCRIPTION Robust-looking, heavily streaked pipit. Brow dirty white with brown ear-coverts. Crown to most of upperparts pale brown. Fine streaking from crown to nape,

but bolder streaks on mantle, scapulars and wing-coverts. Underparts warm brown, with thick streaks on breast sides, but thinner towards flanks. Bill thick with yellowish base. **DISTRIBUTION** Discontinuously Himalayas to SW, S and SE China. Widespread SE China. **HABITS AND HABITATS** Uncommon to locally common resident of grassy hillsides and scrub, mostly afrom 500m to 1,400m. Descends lower in winter. Calls a soft *pit*, while song is a series of *wee-hee*, the second note higher; recalls squeaking sound of a metal hinge. **TIMING** All year round. **SITES** Kanghe (Guangdong), Kuocangshan (Zhejiang). **CONSERVATION** Least Concern.

Richard's Pipit ■ *Anthus richardi* 理氏鹨 (Lǐ shì liù) 18cm

DESCRIPTION ssp. *sinensis* (shown), *ussuriensis*. Large, lanky-looking pipit with an upright posture. Upperparts tawny-brown, streaked on crown, mantle and scapulars. Underparts buff with dark brown streaking confined to breast. Other pipits are more extensively streaked. **DISTRIBUTION** Breeds C Asia, Mongolia, east to E Russia, N, C, E and SE China. Winters S, SE China, Southeast Asia and Indian subcontinent. Widespread across SE China. **HABITS AND HABITATS** Common resident and winter visitor, occurring in open areas like scrub, grassland and farmland. Forages on the ground for insects, singly or in pairs. Sings a series of *kwee-ee* notes. Calls a harsh *cheep* note. **TIMING** All year round. **SITES** Suitable habitat across SE China. **CONSERVATION** Least Concern.

Olive Tree Pipit ■ *Anthus hodgsoni* 树鹨 (Shù liù) 16cm

DESCRIPTION ssp. *yunnanensis*, *hodgsoni* (shown). Small, heavily streaked pipit. Upperparts olive-brown; underparts off-white with bold streaks from breast to flanks and belly. Pale brow followed by a small white spot over the ear coverts. Also fine streaking on back and crown; two wing-bars. **DISTRIBUTION** Breeds W Russia, C Asia, eastwards to Russian Far East, NE China, Korea and Japan, south to SW China and Himalayas. Winters S, SE China and Southeast Asiaand Indian subcontinent. **HABITS AND HABITATS** Common passage migrant and winter visitor, occurring in most kinds of forest, woodland, scrub, farmland and even parks from lowlands to 1,850m. Seen singly, sometimes in small groups. Calls a high-pitched, shrill *pseeet*. **TIMING** Oct–Apr. **SITES** Suitable habitat across SE China. **CONSERVATION** Least Concern.

Pechora Pipit ▪ *Anthus gustavi* 北鹨 (Běi liù) 15cm

DESCRIPTION ssp. *gustavi* (shown), *menzbieri*. Darkest of the streaked pipits. Upperparts brown with strong streaking on mantle (unlike in **Olive-backed Pipit**); also streaked

from breast to lower flanks, but not as boldly as **Olive-backed Pipit**. Darker than **Red-throated Pipit**, and shows black lore patch. Long primaries extend well beyond tertials. **DISTRIBUTION** Breeds N (Urals) to E Russia, NE China and Russian Far East. Winters in Philippines, Borneo and Wallacea. **HABITS AND HABITATS** Uncommon passage migrant, occurring in forests, forest edges, farmland, woodland and wet grassy areas. Usually seen along quiet forest trails. Calls a *pwit* when flushed, but generally silent. **TIMING** Spring passage Apr–May; autumn passage Oct–early Nov. **SITES** Nanhui Dongtan (Shanghai), Hangzhou parks (Zhejiang). **CONSERVATION** Least Concern. Dependent largely on forests in wintering grounds.

Grey-capped Greenfinch ▪ *Chloris sinica* 金翅雀 (Jīn chì què) 14cm

DESCRIPTION ssp. *sinica*. Small, colourful finch. Head grey with greenish-yellow wash on face and chin; bill pale pinkish. Mantle and wing-coverts brown; flight feathers black

with a large yellow patch. Breast to belly washed rufous. Uppertail-coverts olive-green, best seen in flight. **DISTRIBUTION** N Southeast Asia, S, SE China to NE China, Russian Far East, Kamchatka and Japan. Northern populations winter Japan, E, SE China, Taiwan. **HABITS AND HABITATS** Common winter visitor and resident, occurring in mixed and coniferous forests, woodland and edges of cultivation, especially paddyfields. Also in scrub and parkland during winter, usually in flocks. Calls include various twittering with a nasal quality, and a drawn *dweeeh*. **TIMING** All year round; most common in winter due to influx of migrants from north. **SITES** Suitable habitat across SE China. **CONSERVATION** Least Concern.

Brown Bullfinch ■ *Pyrrhula nipalensis* 褐灰雀 (Hè huī què) 16cm

DESCRIPTION ssp. *ricketti*. Medium-sized greyish finch. Forehead and chin black, with a thin white cheek-patch. Head, mantle and breast greyish-brown, graduating to paler brown on belly. Wing-coverts greyish-white, contrasting with glossy black flight feathers. Tail long, black and slightly notched. Young birds lack facial patterns and are browner than adults. **DISTRIBUTION** Himalayas, NE India, mainland Southeast Asia to Malay Peninsula; also SW, S, SE China, including Taiwan. **HABITS AND HABITATS** Uncommon to fairly common resident of broadleaved evergreen, mixed and coniferous forests and forest edges, mostly from 1,400 to 1,850m. Descends to lower elevations in winter. Occurs in pairs or small groups, usually keeping to canopy. **TIMING** All year round. **SITES** Hupingshan (Hunan), Wuyishan, Yangjifeng (Jiangxi). **CONSERVATION** Least Concern.

Chinese Grosbeak ■ *Eophona migratoria* 黑尾蜡嘴雀 (Hēi wěi là zuǐ què) 17cm

DESCRIPTION ssp. *migratoria*. Robust-looking finch with a chunky yellow bill. Male is black headed with a thick yellow bill; rest of upperparts pale greyish-brown; wings glossy black with white tips to flight feathers. Males of similar **Japanese Grosbeak** shows less extensive black on head, which does not extend to ear-coverts; also lack white wing-tips. Female browner than male and lacks black cap. **DISTRIBUTION** Breeds C, E China to NE China, Russian Far East, Korea. Winters Japan, E, S, SE China to N Southeast Asia. **HABITS AND HABITATS** Common resident and winter visitor (except Hainan where rare), occurring in mixed and deciduous forests, forest edges and woodland; also orchards, wooded farms and parks in winter. Forms large flocks, gathering to feed at pine trees. Calls a series of rich whistles, *chwee-chi-chew-chee-wee*. **TIMING** All year round; most common during winter months due to influx of migrants. **SITES** Guangzhou parks (Guangdong), Century Park (Shanghai), Hangzhou parks (Zhejiang). **CONSERVATION** Least Concern.

♀

Common Rosefinch ■ *Carpodacus erythrinus* 普通朱雀
(Pǔ tōng zhū què) 14cm

DESCRIPTION ssp. *grebnitskii, roseatus*. Only regular rosefinch in region. Male unmistakable, with reddish head to breast, and reddish rump; mantle and wing-coverts brownish-grey with a red wash; underparts washed pink. Female has dull greyish-brown upperparts, and streaked pale underparts. **DISTRIBUTION** Breeds W Europe, C Asia east

to NE China, Kamchatka; also Himalayas and Tibetan Plateau. Winters Indian subcontinent, mainland Southeast Asia, S and SE China. **HABITS AND HABITATS** Uncommon to common winter visitor and passage migrant (except Hainan), occurring in

woodland, scrub and farmland from hills to over 2,000m. Usually in singles or small flocks, but sometimes gathers in large flocks at fruiting or flowering trees. **TIMING** Mostly Oct–Apr. **SITES** Suitable habitat across SE China. **CONSERVATION** Least Concern.

Crested Bunting ■ *Melophus lathami* 凤头鹀 (Fèng tóu wū) 17cm

DESCRIPTION Only bunting with a long crest. Male (shown) entirely glossy black with rich chestnut wings; tail chestnut with black tips. Female has shorter crest;

plumage mostly greyish-brown on upperparts; wings washed chestnut; underparts buff with brown streaking. **DISTRIBUTION** Indian subcontinent, Himalayas, N Southeast Asia to SW, S, SE China. Widespread in SE China, from Guangdong to Zhejiang. **HABITS AND HABITATS** Uncommon to rare resident, occurring on scrubby and grassy hillsides, as well as on farmland and in woodland edges, usually above 500m. Usually in pairs. Calls a wheezy, high-pitched series of *ch-chwee-chwee-peeu*, often from a perch. **TIMING** All year round. **SITES** Wuyishan (Jiangxi), Kuocangshan, Xitianmushan (Zhejiang). **CONSERVATION** Least Concern. Has declined greatly in Hong Kong.

Slaty Bunting ▪ *Latoucheornis siemsseni* 蓝鹀 (Lán wū) 13cm

DESCRIPTION Small dark bunting with a shortish, sharp bill. Male dark slaty-blue on most of plumage, with white from belly to undertail-coverts. Female rich orange-chestnut from head to breast, and mantle; wings dark brown with pale edging to feathers. **DISTRIBUTION** C, E, SE China. In SE China, Guangdong to Zhejiang. Endemic to China. **HABITS AND HABITATS** In region, summer breeder in W, S Hunan and possibly W Guangdong, occurring in broadleaved evergreen and mixed forests mostly above 1,200m. Winters in forest edges, woodland and scrub,

mostly in lowlands. Calls a series of high-pitched *psii* notes. Also a soft *tsip*. **TIMING** Breeds Apr–July; winters Nov–Mar. **SITES** Guanshan (Jiangxi), Hupingshan, Mangshan (Hunan), Longwangshan (Zhejiang). **CONSERVATION** Least Concern.

Chestnut Bunting ▪ *Emberiza rutila* 栗鹀 (Lì wū) 13.5cm

DESCRIPTION Richly coloured bunting. Male bright chestnut on head and upperparts; underparts mostly yellow. Female similar to female **Yellow-breasted Bunting**, but has weaker facial patterns and a rich chestnut rump. **DISTRIBUTION** Breeds E, C Siberia to Russian Far East and NE China. Winters mainland Southeast Asia, S and SE China. **HABITS AND HABITATS** Uncommon to fairly common passage migrant and winter visitor, occurring in forest edges, woodland, farmland and open scrub, and on grassy hillsides, from lowlands to over 2,000m. Feeds on the ground, hiding in dense vegetation when disturbed. Usually in small groups, but large

flocks recorded during migration. **TIMING** Oct–Apr. **SITES** Chebaling (Guangdong), Jiulianshan (Jiangxi), Xiaoyangshan (Zhejiang). **CONSERVATION** Least Concern. Surveys in markets showed that this species is heavily trapped.

Tristram's Bunting ▪ *Emberiza tristrami* 白眉鹀 (Bái méi wū) 14.5cm

DESCRIPTION Small bunting with strong facial patterns. Breeding male is black on head and throat, with a white crown-stripe, brow and malar extending to neck sides; white patch over ear-coverts. In non-breeding males face is greyish-brown, not black. Rest of upperparts greyish-brown with dark streaking; flight feathers, uppertail-coverts and tail rich chestnut. Female similar to male but less strongly marked. **DISTRIBUTION** Breeds NE China, Korea and Russian Far East. Winters mostly E, SE China. **HABITS AND HABITATS** Uncommon to fairly common winter visitor, occurring in broadleaved evergreen and mixed forests, woodland and orchards, from hills to nearly 2,000m. Prefers

more wooded areas than other buntings, often quietly feeding on forest floor singly or in small groups. Calls a soft *tsip*. **TIMING** Nov–Mar. **SITES** Tong'an Xiaoping, Fuzhou forest parks (Fujian), Nanling (Guangdong), Hangzhou Botanical Gardens (Zhejiang). **CONSERVATION** Least Concern.

Little Bunting
▪ *Emberiza pusilla* 小鹀 (Xiǎo wū) 13cm

DESCRIPTION Small bunting with a short, petit bill. Breeding male dark-crowned while lores, brow and cheeks rufous, giving a 'rufous-faced' appearance. Mantle and neck sides greyish-brown with dark streaking; wings chestnut-brown with pale edging to feathers. Non-breeding birds (shown) less strongly marked, especially on face, and appear duller. **DISTRIBUTION** Breeds Fennoscandia, W Russia east to Mongolia, NE China, Chukotka and Yakutia. Winters NE India, N Southeast Asia to SW, S, SE China, including Taiwan. **HABITS AND HABITATS** Common winter visitor, occurring mostly in scrub, woodland, orchards and farmland, especially dry grassy and shrubby margins, from lowlands to 2,000m. Shy, feeding in dense areas, and easily flushed. **TIMING** Oct–Apr. **SITES** Century Park, Nanhui Dongtan (Shanghai), Hangzhou parks (Zhejiang). **CONSERVATION** Least Concern.

Yellow-browed Bunting ■ *Emberiza chrysophrys* 黄眉鹀
(Huáng méi wū) 15cm

DESCRIPTION Medium-sized bunting with a slight crest. Breeding male black-capped with thick yellow brow and lores; throat white with thin black malar. Upperparts greyish-brown with dark streaking; chestnut washed on mantle, scapulars and flight feathers; underparts dirty white streaked mainly along flanks. Female (shown) and young birds brownish on cheeks, not black. **DISTRIBUTION** Breeds mostly Transbaikalia to E Russia (Yakutia). Winters S, E, SE China. **HABITS AND HABITATS** Uncommon to locally common winter visitor to mixed forests, forest edges and woodland; sometimes well-wooded fringes of farmland, from hills to 2,000m. Forages with other buntings. Call a thin, ringing *tzeep*. **TIMING** Mostly during passage in Oct and Apr; smaller numbers Oct–Mar. **SITES** Wuyuan (Jiangxi), Nanhui Dongtan (Shanghai), Hangzhou parks (Zhejiang). **CONSERVATION** Least Concern.

Yellow-throated Bunting ■ *Emberiza elegans* 黄喉鹀
(Huáng hóu wū) 15.5cm

DESCRIPTION ssp. *elegans* (shown), *elegantula* (Hunan). Medium-sized, crested bunting with bold facial patterns. Breeding male (shown) black masked, contrasting with rich yellow brow extending to nape, and yellow throat; long black crest and breast-band. Rest of upperparts and wings chestnut with dark streaks. Female shorter crested, dull yellow on face and lacks breast-band. **DISTRIBUTION** Breeds NE China, Russian Far East, Korea; also C, SW China. Northern populations winter Korea, E, SE China, including Taiwan. **HABITS AND HABITATS** Common winter visitor (except Hainan and Guangdong, where rare), occurring at forest edges, in woodland, and on scrubby hillsides and farmland, usually in grassy areas; occasionally in urban parks. Summer breeder in W Hunan, in montane mixed deciduous woodland. Occurs singly or in small groups. Song a rich series of high-pitched, varied notes. **TIMING** Mostly mid-Nov–Apr; breeds May–Jul. **SITES** Chongming Dongtan, Nanhui Dongtan (Shanghai), Hupingshan (Hunan), Hangzhou Botanical Gardens (Zhejiang). **CONSERVATION** Least Concern.

Yellow-breasted Bunting ■ *Emberiza aureola* 黄胸鹀

(Huáng xiōng wū) 15cm

DESCRIPTION ssp. *ornata, aureola*. Medium-sized bunting with yellow underparts. Male unmistakable, with black hood, rich chestnut-brown from nape to much of upperparts, and yellow underparts broken by brown breast-band. Female well marked on face, with dull yellow brow and throat against dark borders of greyish-brown ear-coverts. Underparts dull yellow with streaking. Similar female **Chestnut Bunting** less strongly marked on face and has rufous uppertail-coverts. **DISTRIBUTION** Breeds Fennoscandia eastwards to Chukotka, Kamchatka, and south to N Mongolia, NE China, Hokkaido. Winters NE

India, mainland Southeast Asia, S, SE China. **HABITS AND HABITATS** Uncommon to rare winter visitor and passage migrant, occurring in open scrub, wet grassland and farmland, including dry paddy fields, mostly in lowlands. Calls a dry-sounding *tzip*. **TIMING** Mostly during passage periods Sep and Apr; smaller numbers Nov–Mar. **SITES** Nanling, South China Botanical Gardens (Guangdong), Nanhui Dongtan (Shanghai). **CONSERVATION** Endangered. Has declined greatly due to widespread and heavy trapping.

Yellow Bunting ■ *Emberiza sulphurata* 硫磺鹀 (Liú huáng wū) 14cm

DESCRIPTION Small yellowish bunting. Male's head to mantle, breast and flanks washed greenish-yellow; black lores and chin; broken, white eye-ring. Mantle and scapulars olive-brown with bold black streaking; wing chestnut with pale edging to coverts, forming two wing-bars. Female very similar to male but lacks black facial markings. **DISTRIBUTION** Breeds Honshu. Winters Luzon, SE China, Taiwan. **HABITS AND HABITATS** Uncommon to rare passage migrant, occurring in woodland edges, open tree

plantations, grassland, scrub and farmland. One of the scarcest buntings in the region. **TIMING** Mostly spring passage Mar–mid-Apr. **SITES** Minjiang Estuary (Fujian), Nanhui Dongtan (Shanghai), Xiaoyangshan (Zhejiang).

CONSERVATION Vulnerable. Threatened by habitat loss and some degree of trapping across wintering range.

Black-faced Bunting ■ *Emberiza spodocephala* 灰头鹀 (Huī tóu wū) 14cm

DESCRIPTION ssp. *spodocephala, sordida*. Small, dark-hooded bunting. Male *sordida* (shown) mostly olive-green on hood, with black lores and chin; upperparts and wings brown with dark streaking, pale edged on coverts, forming two wing-bars; underparts yellow with brown streaking on flanks. Female duller with greyish-brown ear-coverts, pale moustache and thin dark malar. In flight note white edges of tail. **DISTRIBUTION** Breeds N Mongolia, Transbaikalia to NE China, Russian Far East, Japan; also C China. Winters NE India, N Southeast Asia, SW, S, SE China to Japan. **HABITS AND HABITATS** Common winter visitor, occurring in mixed and deciduous forests, forest edges, woodland, wet grassland, scrub and farmland, especially in scrubby fringes and near water. One of the most common wintering buntings in region. Calls a metallic, ringing *tsip*. **TIMING** Oct–Apr. **SITES** Tong'an Xiaoping Forest Park (Fujian), Wuyuan, Poyang Lake (Jiangxi), Hangzhou parks (Zhejiang). **CONSERVATION** Least Concern. Have suffered massive decline in Guangdong due to trappping.

Pallas's Bunting ■ *Emberiza pallasi* 苇鹀 (Wěi wū) 13cm

DESCRIPTION ssp. *polaris, minor, lydiae*. Small bunting with a petit bill. Breeding male has a black head broken by a thin white malar; underparts white; upperparts pale brown with bold streaking on mantle and scapulars. Non-breeding male is duller brown on hood, but facial patterns similar. Female brown on ear-coverts, with white malar and buff brow; upperparts streaked. In all forms, lesser wing-coverts greyish, unlike rufous of larger **Common Reed Bunting**. **DISTRIBUTION** Breeds C Siberia, C Asia, Mongolia eastwards to Chukotka, Kamchatka. Winters NE, E, SE China, Korea and Japan. **HABITS AND HABITATS** Uncommon to locally common winter visitor (rare or absent in southern part of region), favouring scrub, dry grassland and farmland with grassy or shrubby fringes, mostly in the lowlands. Usually seen in small flocks. Calls a *chzeep*, recalling tree sparrow calls. **TIMING** Nov–Mar. **SITES** Nanhui Dongtan, Chongming Dongtan (Shanghai). **CONSERVATION** Least Concern.

polaris　　　　*lydiae*

Chinese and English names follows the CBR Checklist to the Birds of China v 3.0 (2013)

DEFINITIONS OF STATUS

Resident: species that occurs all year round within region

Resident (introduced): species that regularly occur in the region due to human assistance, and show evidence of breeding

Summer visitor: species that breeds in the region during spring-summer months, but overwinter south of the region

Winter visitor: species not known to breed in the region, but known to occur regularly during winter months

Passage migrant: species occuring only during the migration period in spring/autumn months, and overwintering south of the region

Non-breeding visitor: wide ranging species from nearby regions that stray into the region, but no evidence of breeding yet exists

Vagrant: species occuring outside of their regular range, but in an apparently wild state

ABBREVIATIONS FOR CONSERVATION STATUS

CR: Critically Endangered, **EN:** Endangered, **VU:** Vulnerable, **NT:** Near Threatened,
LC: Least Concern, **NE:** Not evaluated

Common name	Scientific name	Chinese name	Status	IUCN Red List Status
Phasianidae (Pheasants and Allies)				
Chinese Francolin	*Francolinus pintadeanus*	中华鹧鸪	Resident	LC
Japanese Quail	*Coturnix japonica*	鹌鹑	Winter visitor/passage migrant	NT
King Quail	*Excalfactoria chinensis*	蓝胸鹑	Resident/non-breeding visitor?	LC
White-necklaced Partridge	*Arborophila gingica*	白眉山鹧鸪	Resident	NT
Hainan Partridge	*Arborophila ardens*	海南山鹧鸪	Resident	VU
Chinese Bamboo Partridge	*Bambusicola thoracicus*	灰胸竹鸡	Resident	LC
Temminck's Tragopan	*Tragopan temminckii*	红腹角雉	Resident	LC
Cabot's Tragopan	*Tragopan caboti*	黄腹角雉	Resident	VU
Koklass Pheasant	*Pucrasia macrolopha*	勺鸡	Resident	LC
Red Junglefowl	*Gallus gallus*	红原鸡	Resident	LC
Silver Pheasant	*Lophura nycthemera*	白鹇	Resident	LC
Elliot's Pheasant	*Syrmaticus ellioti*	白颈长尾雉	Resident	NT
Reeves's Pheasant	*Syrmaticus reevesii*	白冠长尾雉	Resident	VU
Common Pheasant	*Phasianus colchicus*	雉鸡	Resident	LC
Golden Pheasant	*Chrysolophus pictus*	红腹锦鸡	Resident	LC
Hainan Peacock Pheasant	*Polyplectron katsumatae*	海南孔雀雉	Resident	EN
Anatidae (Ducks and Geese)				
Lesser Whistling Duck	*Dendrocygna javanica*	栗树鸭	Resident	LC
Swan Goose	*Anser cygnoides*	鸿雁	Winter visitor	VU
Taiga Bean Goose	*Anser fabalis*	豆雁	Winter visitor	LC
Tundra Bean Goose	*Anser serrirostris*	短嘴豆雁	Winter visitor	LC
Greylag Goose	*Anser anser*	灰雁	Winter visitor	LC
Greater White-fronted Goose	*Anser albifrons*	白额雁	Winter visitor	LC
Lesser White-fronted Goose	*Anser erythropus*	小白额雁	Winter visitor	VU
Bar-headed Goose	*Anser indicus*	斑头雁	Vagrant	LC
Snow Goose	*Chen caerulescens*	雪雁	Vagrant	LC
Brant Goose	*Branta bernicla*	黑雁	Vagrant	LC
Red-breasted Goose	*Branta ruficollis*	红胸黑雁	Vagrant	EN
Tundra Swan	*Cygnus columbianus*	鸿雁	Winter visitor	LC
Common Shelduck	*Tadorna tadorna*	翘鼻麻鸭	Winter visitor	LC
Ruddy Shelduck	*Tadorna ferruginea*	赤麻鸭	Winter visitor	LC
Mandarin Duck	*Aix galericulata*	鸳鸯	Winter visitor	LC
Cotton Pygmy Goose	*Nettapus coromandelianus*	棉凫	Summer visitor	LC
Gadwall	*Anas strepera*	赤膀鸭	Winter visitor	LC
Falcated Duck	*Anas falcata*	罗纹鸭	Winter visitor	LC
Eurasian Wigeon	*Anas penelope*	赤颈鸭	Winter visitor	LC
Mallard	*Anas platyrhynchos*	绿头鸭	Winter visitor	LC

Common name	Scientific name	Chinese name	Status	IUCN Red List Status
Eastern Spot-billed Duck	*Anas zonorhyncha*	斑嘴鸭	Resident	LC
Northern Shoveller	*Anas clypeata*	琵嘴鸭	Winter visitor	LC
Northern Pintail	*Anas acuta*	针尾鸭	Winter visitor	LC
Garganey	*Anas querquedula*	白眉鸭	Winter visitor	LC
Baikal Teal	*Anas formosa*	花脸鸭	Winter visitor	LC
Eurasian Teal	*Anas crecca*	绿翅鸭	Winter visitor	LC
Red-crested Pochard	*Netta rufina*	赤嘴潜鸭	Winter visitor	LC
Common Pochard	*Aythya ferina*	红头潜鸭	Winter visitor	LC
Baer's Pochard	*Aythya baeri*	青头潜鸭	Winter visitor	CR
Ferruginous Duck	*Aythya nyroca*	白眼潜鸭	Vagrant	NT
Tufted Duck	*Aythya fuligula*	凤头潜鸭	Winter visitor	LC
Greater Scaup	*Aythya marila*	斑背潜鸭	Winter visitor	LC
Black Scoter	*Melanitta americana*	黑海番鸭	Vagrant	NT
White-winged Scoter	*Melanitta deglandi*	斑脸海番鸭	Winter visitor	LC
Long-tailed Duck	*Clangula hyemalis*	长尾鸭	Vagrant	VU
Common Goldeneye	*Bucephala clangula*	鹊鸭	Winter visitor	LC
Smew	*Mergellus albellus*	白秋沙鸭	Winter visitor	LC
Common Merganser	*Mergus merganser*	普通秋沙鸭	Winter visitor	LC
Red-breasted Merganser	*Mergus serrator*	红胸秋沙鸭	Winter visitor	LC
Scaly-sided Merganser	*Mergus squamatus*	中华秋沙鸭	Winter visitor	EN
Gaviidae (Loons)				
Red-throated Loon	*Gavia stellata*	红喉潜鸟	Winter visitor	LC
Black-throated Loon	*Gavia arctica*	黑喉潜鸟	Winter visitor	LC
White-billed Loon	*Gavia adamsii*	黄嘴潜鸟	Vagrant	NT
Diomedeidae (Albatrosses)				
Black-footed Albatross	*Phoebastria nigripes*	黑脚信天翁	Vagrant	NT
Short-tailed Albatross	*Phoebastria albatrus*	短尾信天翁	Vagrant	VU
Procellariidae (Petrels and Shearwaters)				
Bonin Petrel	*Pterodroma hypoleuca*	白额圆尾鹱	Vagrant	LC
Streaked Shearwater	*Calonectris leucomelas*	白额鹱	Passage migrant	LC
Short-tailed Shearwater	*Puffinus tenuirostris*	短尾鹱	Passage migrant	LC
Wedge-tailed Shearwater	*Puffinus pacificus*	楔尾鹱	Passage migrant	LC
Sooty Shearwater	*Puffinus griseus*	灰鹱	Non-breeding visitor	LC
Bulwer's Petrel	*Bulweria bulwerii*	褐燕鹱	Summer visitor	LC
Hydrobatidae (Storm Petrels)				
Swinhoe's Storm Petrel	*Oceanodroma monorhis*	黑叉尾海燕	Summer visitor/passage migrant	NT
Podicipedidae (Grebes)				
Little Grebe	*Tachybaptus ruficollis*	小鸊鷉	Resident	LC
Red-necked Grebe	*Podiceps grisegena*	赤颈鸊鷉	Winter visitor	LC
Great Crested Grebe	*Podiceps cristatus*	凤头鸊鷉	Winter visitor	LC
Horned Grebe	*Podiceps auritus*	角鸊鷉	Winter visitor	LC
Black-necked Grebe	*Podiceps nigrocollis*	黑颈鸊鷉	Winter visitor	LC
Ciconiidae (Storks)				
Painted Stork	*Mycteria leucocephala*	彩鹳	Vagrant	NT
Asian Openbill	*Anastomus oscitans*	钳嘴鹳	Non-breeding visitor	LC
Black Stork	*Ciconia nigra*	黑鹳	Winter visitor	LC
Oriental Stork	*Ciconia boyciana*	东方白鹳	Winter visitor/resident	EN
Threskiornithidae (Ibises, Spoonbills)				
Black-headed Ibis	*Threskiornis melanocephala*	黑头白鹮	Non-breeding visitor	NT
Glossy Ibis	*Plegadis falcinellus*	彩鹮	Vagrant	LC
Eurasian Spoonbill	*Platalea leucorodia*	白琵鹭	Winter visitor	LC
Black-faced Spoonbill	*Platalea minor*	黑脸琵鹭	Winter visitor	EN
Ardeidae (Herons and Bitterns)				
Great Bittern	*Botaurus stellaris*	大麻鳽	Winter visitor	LC
Yellow Bittern	*Ixobrychus sinensis*	黄苇鳽	Summer visitor/passage migrant/winter visitor	LC
Von Schrenk's Bittern	*Ixobrychus eurhythmus*	紫背苇鳽	Passage migrant	LC
Cinnamon Bittern	*Ixobrychus cinnamomeus*	栗苇鳽	Resident	LC
Black Bittern	*Dupetor flavicollis*	黑鳽	Summer visitor/passage migrant	LC
White-eared Night Heron	*Gorsachius magnificus*	海南鳽	Resident	EN
Japanese Night Heron	*Gorsachius goisagi*	栗鳽	Vagrant	EN
Malayan Night Heron	*Gorsachius melanolophus*	黑冠鳽	Summer visitor	LC
Black-crowned Night Heron	*Nycticorax nycticorax*	夜鹭	Resident/summer visitor/winter visitor	LC

Common name	Scientific name	Chinese name	Status	IUCN Red List Status
Striated Heron	*Butorides striata*	绿鹭	Resident/summer visitor/winter visitor	LC
Chinese Pond Heron	*Ardeola bacchus*	池鹭	Resident/summer visitor/winter visitor	LC
Eastern Cattle Egret	*Bubulcus coromandus*	牛背鹭	Resident/summer visitor/winter visitor	LC
Grey Heron	*Ardea cinerea*	苍鹭	Winter visitor/resident	LC
Purple Heron	*Ardea purpurea*	草鹭	Resident/winter visitor	LC
Great Egret	*Ardea alba*	大白鹭	Resident/summer visitor/winter visitor	LC
Intermediate Egret	*Egretta intermedia*	中白鹭	Resident/winter visitor	LC
Little Egret	*Egretta garzetta*	白鹭	Resident/summer visitor/winter visitor	LC
Pacific Reef Heron	*Egretta sacra*	岩鹭	Resident	LC
Chinese Egret	*Egretta eulophotes*	黄嘴白鹭	Summer visitor/passage migrant	VU
Pelecanidae (Pelicans)				
Spot-billed Pelican	*Pelecanus philippensis*	斑嘴鹈鹕	Vagrant	NT
Dalmatian Pelican	*Pelecanus crispus*	卷羽鹈鹕	Winter visitor	VU
Fregatidae (Frigatebirds)				
Great Frigatebird	*Fregata minor*	小军舰鸟	Non-breeding visitor	LC
Lesser Frigatebird	*Fregata ariel*	白斑军舰鸟	Non-breeding visitor	LC
Christmas Frigatebird	*Fregata andrewsi*	白腹军舰鸟	Non-breeding visitor	CR
Sulidae (Gannets and Boobies)				
Red-footed Booby	*Sula sula*	红脚鲣鸟	Non-breeding visitor	LC
Brown Booby	*Sula leucogaster*	褐鲣鸟	Non-breeding visitor	LC
Phalacrocoracidae (Cormorants)				
Pelagic Cormorant	*Phalacrocorax pelagicus*	海鸬鹚	Non-breeding visitor	LC
Great Cormorant	*Phalacrocorax carbo*	普通鸬鹚	Winter visitor	LC
Japanese Cormorant	*Phalacrocorax capillatus*	暗绿背鸬鹚	Winter visitor	LC
Pandionidae (Ospreys)				
Western Osprey	*Pandion haliaeetus*	鹗	Winter visitor/passage migrant/ non-breeding visitor	LC
Accipitridae (Kites, Hawks and Eagles)				
Black-winged Kite	*Elanus caeruleus*	黑翅鸢	Resident	LC
Crested Honey Buzzard	*Pernis ptilorhynchus*	凤头蜂鹰	Summer visitor/passage migrant/ winter visitor	LC
Black Baza	*Aviceda leuphotes*	黑冠鹃隼	Summer visitor/passage migrant	LC
Cinereous Vulture	*Aegypius monachus*	秃鹫	Winter visitor	NT
Crested Serpent Eagle	*Spilornis cheela*	蛇雕	Resident	LC
Mountain Hawk-Eagle	*Nisaetus nipalensis*	鹰雕	Resident	LC
Black Eagle	*Ictinaetus malayensis*	林雕	Resident	LC
Greater Spotted Eagle	*Clanga clanga*	乌雕	Winter visitor	VU
Steppe Eagle	*Aquila nipalensis*	草原雕	Vagrant	LC
Eastern Imperial Eagle	*Aquila heliaca*	白肩雕	Winter visitor	VU
Golden Eagle	*Aquila chrysaetos*	金雕	Non-breeding visitor	LC
Bonelli's Eagle	*Aquila fasciatus*	白腹隼雕	Resident	LC
Crested Goshawk	*Accipiter trivirgatus*	凤头鹰	Resident	LC
Shikra	*Accipiter badius*	褐耳鹰	Resident	LC
Chinese Sparrowhawk	*Accipiter soloensis*	赤腹鹰	Summer visitor/passage migrant	LC
Japanese Sparrowhawk	*Accipiter gularis*	日本松雀鹰	Passage migrant/winter visitor	LC
Besra	*Accipiter virgatus*	松雀鹰	Resident	LC
Eurasian Sparrowhawk	*Accipiter nisus*	雀鹰	Winter visitor/passage migrant	LC
Northern Goshawk	*Accipiter gentilis*	苍鹰	Winter visitor	LC
Eastern Marsh Harrier	*Circus spilonotus*	白腹鹞	Winter visitor	LC
Hen Harrier	*Circus cyaneus*	白尾鹞	Winter visitor	LC
Pied Harrier	*Circus melanoleucos*	鹊鹞	Passage migrant/winter visitor	LC
Black Kite	*Milvus migrans*	黑鸢	Resident/summer visitor/winter visitor	LC
Brahminy Kite	*Haliastur indus*	栗鸢	Non-breeding visitor	LC
White-bellied Sea Eagle	*Haliaeetus leucogaster*	白腹海雕	Resident	LC
Lesser Fish Eagle	*Haliaeetus humilis*	渔雕	Non-breeding visitor	NT
Pallas's Fish Eagle	*Haliaeetus leucoryphus*	玉带海雕	Vagrant	VU
White-tailed Sea Eagle	*Haliaeetus albicilla*	白尾海雕	Winter visitor	NT
Grey-faced Buzzard	*Butastur indicus*	灰脸鵟鹰	Passage migrant	LC
Rough-legged Buzzard	*Buteo lagopus*	毛脚鵟	Winter visitor	LC

Common name	Scientific name	Chinese name	Status	IUCN Red List Status
Eastern Buzzard	Buteo japonicus	普通鵟	Winter visitor	LC
Upland Buzzard	Buteo hemilasius	大鵟	Winter visitor	LC
Falconidae (Falcons)				
Pied Falconet	Microhierax melanoleucos	白腿小隼	Resident	LC
Common Kestrel	Falco tinnunculus	红隼	Resident/winter visitor	LC
Amur Falcon	Falco amurensis	红脚隼	Passage migrant	LC
Merlin	Falco columbarius	灰背隼	Winter visitor	LC
Eurasian Hobby	Falco subbuteo	燕隼	Summer visitor/passage migrant	LC
Peregrine Falcon	Falco peregrinus	游隼	Resident/winter visitor	LC
Otididae (Bustards)				
Great Bustard	Otis tarda	大鸨	Winter visitor	VU
Rallidae (Rails, Crakes and Coots)				
Swinhoe's Rail	Coturnicops exquisitus	花田鸡	Passage migrant/winter visitor	VU
Slaty-legged Crake	Rallina eurizonoides	白喉斑秧鸡	Summer visitor/passage migrant/winter visitor	LC
Slaty-breasted Rail	Gallirallus striatus	蓝胸秧鸡	Resident	LC
Brown-cheeked Rail	Rallus indicus	普通秧鸡	Winter visitor	LC
Brown Crake	Amaurornis akool	红脚苦恶鸟	Resident	LC
White-breasted Waterhen	Amaurornis phoenicurus	白胸苦恶鸟	Resident/passage migrant	LC
Baillon's Crake	Porzana pusilla	小田鸡	Passage migrant	LC
Ruddy-breasted Crake	Porzana fusca	红胸田鸡	Summer visitor/passage migrant/winter visitor	LC
Band-bellied Crake	Porzana paykullii	斑胁田鸡	Passage migrant	NT
Watercock	Gallicrex cinerea	董鸡	Summer visitor	LC
Purple Swamphen	Porphyrio porphyrio	紫水鸡	Resident	LC
Common Moorhen	Gallinula chloropus	黑水鸡	Resident/winter visitor	LC
Eurasian Coot	Fulica atra	骨顶鸡	Winter visitor	LC
Gruidae (Cranes)				
Siberian Crane	Grus leucogeranus	白鹤	Winter visitor	CR
White-naped Crane	Grus vipio	白枕鹤	Winter visitor	VU
Sandhill Crane	Grus canadensis	沙丘鹤	Vagrant	LC
Common Crane	Grus grus	灰鹤	Winter visitor	LC
Hooded Crane	Grus monacha	白头鹤	Winter visitor	VU
Turnicidae (Buttonquails)				
Common Buttonquail	Turnix sylvaticus	林三趾鹑	Resident	LC
Yellow-legged Buttonquail	Turnix tanki	黄脚三趾鹑	Resident /winter visitor/passage migrant	LC
Barred Buttonquail	Turnix suscitator	棕三趾鹑	Resident/passage migrant	LC
Haematopodidae (Oystercatchers)				
Eurasian Oystercatcher	Haematopus ostralegus	蛎鹬	Winter visitor	LC
Recurvirostridae (Stilts, Avocets)				
Black-winged Stilt	Himantopus himantopus	黑翅长脚鹬	Winter visitor/passage migrant/summer visitor	LC
Pied Avocet	Recurvirostra avosetta	反嘴鹬	Winter visitor	LC
Charadriidae (Lapwings, Plovers)				
Northern Lapwing	Vanellus vanellus	凤头麦鸡	Winter visitor	LC
Grey-headed Lapwing	Vanellus cinereus	灰头麦鸡	Summer visitor/passage migrant/winter visitor	LC
Pacific Golden Plover	Pluvialis fulva	金斑鸻	Winter visitor/passage migrant	LC
Grey Plover	Pluvialis squatarola	灰斑鸻	Winter visitor/passage migrant	LC
Common Ringed Plover	Charadrius hiaticula	剑鸻	Winter visitor	LC
Long-billed Plover	Charadrius placidus	长嘴剑鸻	Resident/winter visitor	LC
Little Ringed Plover	Charadrius dubius	金眶鸻	Resident/summer visitor/winter visitor	LC
Kentish Plover	Charadrius alexandrinus	环颈鸻	Summer visitor/passage migrant/winter visitor	LC
Lesser Sand Plover	Charadrius mongolus	蒙古沙鸻	Passage migrant/winter visitor	LC
Greater Sand Plover	Charadrius leschenaultii	铁嘴沙鸻	Passage migrant/winter visitor	LC
Oriental Plover	Charadrius veredus	东方鸻	Passage migrant	LC
Rostratulidae (Painted Snipes)				
Greater Painted Snipe	Rostratula benghalensis	彩鹬	Resident/passage migrant/winter visitor	LC
Jacanidae (Jacanas)				
Pheasant-tailed Jacana	Hydrophasianus chirurgus	水雉	Summer visitor/passage migrant	LC

Common name	Scientific name	Chinese name	Status	IUCN Red List Status
Scolopacidae (Sandpipers and Snipes)				
Eurasian Woodcock	Scolopax rusticola	丘鹬	Winter visitor/passage migrant	LC
Jack Snipe	Lymnocryptes minimus	姬鹬	Vagrant	LC
Latham's Snipe	Gallinago hardwickii	澳南沙锥	Vagrant	LC
Solitary Snipe	Gallinago solitaria	孤沙锥	Vagrant	LC
Pin-tailed Snipe	Gallinago stenura	针尾沙锥	Passage migrant/winter visitor	LC
Swinhoe's Snipe	Gallinago megala	大沙锥	Winter visitor/passage migrant	LC
Common Snipe	Gallinago gallinago	扇尾沙锥	Winter visitor/passage migrant	LC
Long-billed Dowitcher	Limnodramus scolopaceus	长嘴鹬	Winter visitor	LC
Asian Dowitcher	Limnodramus semipalmatus	半蹼鹬	Passage migrant	NT
Black-tailed Godwit	Limosa limosa	黑尾塍鹬	Winter visitor/passage migrant	NT
Bar-tailed Godwit	Limosa lapponica	斑尾塍鹬	Winter visitor/passage migrant	LC
Little Curlew	Numenius minutus	小杓鹬	Passage migrant	LC
Whimbrel	Numenius phaeopus	中杓鹬	Passage migrant/winter visitor	LC
Eurasian Curlew	Numenius arquata	白腰杓鹬	Winter visitor/passage migrant	NT
Far Eastern Curlew	Numenius madagascariensis	大杓鹬	Passage migrant	VU
Spotted Redshank	Tringa erythropus	鹤鹬	Winter visitor	LC
Common Redshank	Tringa totanus	红脚鹬	Winter visitor/passage migrant	LC
Marsh Sandpiper	Tringa stagnatilis	泽鹬	Winter visitor/passage migrant	LC
Common Greenshank	Tringa nebularia	青脚鹬	Winter visitor/passage migrant	LC
Nordmann's Greenshank	Tringa guttifer	小青脚鹬	Passage migrant	EN
Green Sandpiper	Tringa ochropus	白腰草鹬	Winter visitor/passage migrant	LC
Wood Sandpiper	Tringa glareola	林鹬	Winter visitor/passage migrant	LC
Grey-tailed Tattler	Tringa brevipes	灰尾漂鹬	Passage migrant	NT
Terek Sandpiper	Xenus cinereus	翘嘴鹬	Passage migrant	LC
Common Sandpiper	Actitis hypoleucos	矶鹬	Winter visitor/passage migrant	LC
Ruddy Turnstone	Arenaria interpres	翻石鹬	Winter visitor/passage migrant	LC
Great Knot	Calidris tenuirostris	大滨鹬	Passage migrant/winter visitor	VU
Red Knot	Calidris canutus	红腹滨鹬	Passage migrant/winter visitor	LC
Sanderling	Calidris alba	三趾滨鹬	Winter visitor/passage migrant	LC
Red-necked Stint	Calidris ruficollis	红颈滨鹬	Passage migrant/winter visitor	LC
Little Stint	Calidris minuta	小滨鹬	Passage migrant	LC
Temminck's Stint	Calidris temminckii	青脚滨鹬	Passage migrant/winter visitor	LC
Long-toed Stint	Calidris subminuta	长趾滨鹬	Passage migrant/winter visitor	LC
Pectoral Sandpiper	Calidris melanotos	斑胸滨鹬	Passage migrant	LC
Sharp-tailed Sandpiper	Calidris acuminata	尖尾滨鹬	Passage migrant	LC
Curlew Sandpiper	Calidris ferruginea	弯嘴滨鹬	Passage migrant	LC
Dunlin	Calidris alpina	黑腹滨鹬	Winter visitor	LC
Spoon-billed Sandpiper	Eurynorhynchus pygmeus	勺嘴鹬	Winter visitor/passage migrant	CR
Broad-billed Sandpiper	Limicola falcinellus	阔嘴鹬	Passage migrant/winter visitor	LC
Ruff	Philomachus pugnax	流苏鹬	Winter visitor/passage migrant	LC
Red-necked Phalarope	Phalaropus lobatus	红颈瓣蹼鹬	Passage migrant	LC
Red Phalarope	Phalaropus fulicarius	灰瓣蹼鹬	Vagrant	LC
Glareolidae (Pratincoles)				
Oriental Pratincole	Glareola maldivarum	普通燕鸻	Summer visitor/passage migrant	LC
Laridae (Gulls, Terns and Skimmers)				
Black-legged Kittiwake	Rissa tridactyla	三趾鸥	Winter visitor	LC
Slender-billed Gull	Chroicocephalus genei	细嘴鸥	Vagrant	LC
Common Black-headed Gull	Chroicocephalus ridibundus	红嘴鸥	Winter visitor	LC
Brown-headed Gull	Chroicocephalus brunnicephalus	棕头鸥	Passage migrant/winter visitor	LC
Saunder's Gull	Chroicocephalus saundersi	黑嘴鸥	Winter visitor	VU
Little Gull	Hydrocoloeus minutus	小鸥	Vagrant	LC
Relict Gull	Ichthyaetus relictus	遗鸥	Winter visitor	VU
Black-tailed Gull	Larus crassirostris	黑尾鸥	Winter visitor/passage migrant/ summer visitor	LC
Mew Gull	Larus canus	海鸥	Winter visitor	LC
Glacous-winged Gull	Larus glaucescens	灰翅鸥	Vagrant	LC
Glaucous Gull	Larus hyperboreus	北极鸥	Vagrant	LC
Vega Gull	Larus vegae	西伯利亚银鸥	Winter visitor	LC
Caspian Gull	Larus cachinnans	黄脚银鸥	Winter visitor	LC
Slaty-backed Gull	Larus schistisagus	灰背鸥	Winter visitor	LC
Heuglin's Gull	Larus heuglini	乌灰银鸥	Winter visitor	LC

Common name	Scientific name	Chinese name	Status	IUCN Red List Status
Gull-billed Tern	*Gelochelidon nilotica*	鸥嘴噪鸥	Passage migrant/non-breeding visitor	LC
Caspian Tern	*Hydroprogne caspia*	红嘴巨鸥	Winter visitor/passage migrant	LC
Greater Crested Tern	*Thalasseus bergii*	大凤头燕鸥	Summer visitor/passage migrant	LC
Lesser Crested Tern	*Thalasseus bengalensis*	小凤头燕鸥	Non-breeding visitor	LC
Chinese Crested Tern	*Thalasseus bernsteini*	中华凤头燕鸥	Summer visitor	CR
Little Tern	*Sternula albifrons*	白额燕鸥	Summer visitor/passage migrant	LC
Aleutian Tern	*Onychoprion aleuticus*	白腰燕鸥	Passage migrant	LC
Bridled Tern	*Onychoprion anaethetus*	褐翅燕鸥	Summer visitor	LC
Sooty Tern	*Onychoprion fuscatus*	乌燕鸥	Non-breeding visitor	LC
Roseate Tern	*Sterna dougallii*	粉红燕鸥	Summer visitor	LC
Black-naped Tern	*Sterna sumatrana*	黑枕燕鸥	Summer visitor	LC
Common Tern	*Sterna hirundo*	普通燕鸥	Passage migrant	LC
Whiskered Tern	*Chlidonias hybrida*	须浮鸥	Summer visitor/passage migrant/ winter visitor	LC
White-winged Tern	*Chlidonias leucopterus*	白翅浮鸥	Passage migrant	LC
Stercorariidae (Skuas)				
Pomarine Jaeger	*Stercorarius pomarinus*	中贼鸥	Passage migrant/winter visitor	LC
Parasitic Jaeger	*Stercorarius parasiticus*	短尾贼鸥	Passage migrant	LC
Long-tailed Jaeger	*Stercorarius longicaudus*	长尾贼鸥	Passage migrant	LC
Alcidae (Auks)				
Long-billed Murrelet	*Brachyramphus perdix*	斑海雀	Non-breeding visitor	NT
Japanese Murrelet	*Synthliboramphus wumizusume*	冠海雀	Vagrant	VU
Ancient Murrelet	*Synthliboramphus antiquus*	扁嘴海雀	Non-breeding visitor	LC
Columbidae (Pigeons and Doves)				
Rock Dove	*Columba livia*	原鸽	Resident (introduced)	LC
Hill Pigeon	*Columba rupestris*	岩鸽	Non-breeding visitor	LC
Pale-capped Pigeon	*Columba punicea*	紫林鸽	Resident	VU
Oriental Turtle Dove	*Streptopelia orientalis*	山斑鸠	Resident/winter visitor	LC
Eurasian Collared Dove	*Steptopelia decaocto*	灰斑鸠	Resident (introduced)/winter visitor?	LC
Red Turtle Dove	*Streptopelia tranquebarica*	火斑鸠	Resident	LC
Spotted Dove	*Spilopelia chinensis*	珠颈斑鸠	Resident	LC
Barred Cuckoo-Dove	*Macropygia unchall*	斑尾鹃鸠	Resident	LC
Common Emerald Dove	*Chalcophaps indica*	绿翅金鸠	Resident	LC
Orange-breasted Green Pigeon	*Treron bicincta*	橙胸绿鸠	Resident	LC
Thick-billed Green Pigeon	*Treron curvirostra*	厚嘴绿鸠	Resident	LC
Wedge-tailed Green Pigeon	*Treron sphenurus*	楔尾绿鸠	Resident	LC
White-bellied Green Pigeon	*Treron sieboldii*	红翅绿鸠	Resident	LC
Green Imperial Pigeon	*Ducula aenea*	绿皇鸠	Resident	LC
Mountain Imperial Pigeon	*Ducula badia*	山皇鸠	Resident	LC
Psittacidae (Parrots)				
Red-breasted Parakeet	*Psittacula alexandri*	绯胸鹦鹉	Resident	NT
Cuculidae (Cuckoos)				
Greater Coucal	*Centropus sinensis*	褐翅鸦鹃	Resident	LC
Lesser Coucal	*Centropus bengalensis*	小鸦鹃	Resident	LC
Green-billed Malkoha	*Phaenicophaeus tristis*	绿嘴地鹃	Resident	LC
Chestnut-winged Cuckoo	*Clamator coromandus*	红翅凤头鹃	Summer visitor	LC
Asian Koel	*Eudynamys scolopacea*	噪鹃	Resident/summer visitor	LC
Asian Emerald Cuckoo	*Chrysococcyx maculatus*	翠金鹃	Resident	LC
Plaintive Cuckoo	*Cacomantis merulinus*	八声杜鹃	Summer visitor/resident	LC
Fork-tailed Drongo Cuckoo	*Surniculus dicruroides*	乌鹃	Summer visitor	LC
Large Hawk-Cuckoo	*Hierococcyx sparverioides*	鹰鹃	Summer visitor/passage migrant	LC
Rufous Hawk-Cuckoo	*Hierococcyx hyperythrus*	北鹰鹃	Passage migrant	LC
Hodgson's Hawk-Cuckoo	*Hierococcyx nisicolor*	霍氏鹰鹃	Summer visitor	LC
Lesser Cuckoo	*Cuculus policephalus*	小杜鹃	Summer visitor/passage migrant	LC
Indian Cuckoo	*Cuculus micropterus*	四声杜鹃	Resident/summer visitor/passage migrant	LC
Himalayan Cuckoo	*Cuculus saturatus*	中杜鹃	Summer visitor	LC
Oriental Cuckoo	*Cuculus optatus*	北方中杜鹃	Passage migrant	LC
Common Cuckoo	*Cuculus canorus*	大杜鹃	Passage migrant	LC
Tytonidae (Barn Owls)				
Eastern Grass Owl	*Tyto longimembris*	草鸮	Resident	LC
Oriental Bay Owl	*Phodilus badius*	栗鸮	Resident	LC

Common name	Scientific name	Chinese name	Status	IUCN Red List Status
Strigidae (Owls)				
Mountain Scops Owl	*Otus spilocephalus*	黄嘴角鸮	Resident	LC
Collared Scops Owl	*Otus lettia*	西领角鸮	Resident	LC
Oriental Scops Owl	*Otus sunia*	红角鸮	Resident/summer visitor/passage migrant	LC
Eurasian Eagle-Owl	*Bubo bubo*	雕鸮	Resident	LC
Spot-bellied Eagle-Owl	*Bubo nipalensis*	林雕鸮	Resident	LC
Brown Fish Owl	*Ketupa zeylonensis*	褐渔鸮	Resident	LC
Tawny Fish Owl	*Ketupa flavipes*	黄腿渔鸮	Resident	LC
Brown Wood Owl	*Strix leucogrammica*	褐林鸮	Resident	LC
Himalayan Owl	*Strix nivicolum*	灰林鸮	Resident	LC
Collared Owlet	*Glaucidium brodiei*	领鸺鹠	Resident	LC
Asian Barred Owlet	*Glaucidium cuculoides*	斑头鸺鹠	Resident	LC
Brown Hawk-Owl	*Ninox scutulata*	鹰鸮	Resident	LC
Northern Boobook	*Ninox japonica*	北鹰鸮	Passage migrant	LC
Long-eared Owl	*Asio otus*	长耳鸮	Winter visitor	LC
Short-eared Owl	*Asio flammeus*	短耳鸮	Winter visitor	LC
Caprimulgidae (Nightjars)				
Grey Nightjar	*Caprimulgus jotaka*	普通夜鹰	Summer visitor/passage migrant	LC
Large-tailed Nightjar	*Caprimulgus macrurus*	长尾夜鹰	Resident	LC
Savanna Nightjar	*Caprimulgus affinis*	林夜鹰	Resident	LC
Apodidae (Swifts)				
Himalayan Swiftlet	*Aerodramus brevirostris*	短嘴金丝燕	Summer visitor/passage migrant	LC
White-throated Needletail	*Hirundapus caudacutus*	白喉针尾雨燕	Passage migrant	LC
Silver-backed Needletail	*Hirundapus cochinchinensis*	灰喉针尾雨燕	Passage migrant/summer visitor	LC
Asian Palm Swift	*Cypsiurus balasiensis*	棕雨燕	Resident	LC
Pacific Swift	*Apus pacificus*	白腰雨燕	Passage migrant/summer visitor/ winter visitor	LC
Salim Ali's Swift	*Apus salimali*	藏雨燕	Resident?	LC
Cook's Swift	*Apus cooki*	白腰雨燕	Summer visitor?	LC
House Swift	*Apus nipalensis*	小白腰雨燕	Resident/passage migrant	LC
Trogonidae (Trogons)				
Red-headed Trogon	*Harpactes erythrocephalus*	红头咬鹃	Resident	LC
Coraciidae (Rollers)				
Common Dollarbird	*Eurystomus orientalis*	三宝鸟	Summer visitor/passage migrant	LC
Alcedinidae (Kingfishers)				
Ruddy Kingfisher	*Halcyon coromanda*	赤翡翠	Vagrant	LC
White-throated Kingfisher	*Halcyon smyrnensis*	白胸翡翠	Resident/winter visitor	LC
Black-capped Kingfisher	*Halcyon pileata*	蓝翡翠	Winter visitor/passage migrant/ summer visitor	LC
Collared Kingfisher	*Todiramphus chloris*	白领翡翠	Vagrant	LC
Common Kingfisher	*Alcedo atthis*	普通翠鸟	Resident	LC
Blyth's Kingfisher	*Alcedo hercules*	斑头大翠鸟	Resident	NT
Oriental Dwarf Kingfisher	*Ceyx erithaca*	三趾翠鸟	Resident	LC
Crested Kingfisher	*Megaceryle lugubris*	冠鱼狗	Resident	LC
Pied Kingfisher	*Ceryle rudis*	斑鱼狗	Resident	LC
Meropidae (Bee-eaters)				
Blue-bearded Bee-eater	*Nyctyornis athertoni*	蓝须夜蜂虎	Resident	LC
Blue-throated Bee-eater	*Merops viridis*	蓝喉蜂虎	Summer visitor	LC
Blue-tailed Bee-eater	*Merops philippinus*	栗喉蜂虎	Summer visitor	LC
Upupidae (Hoopoes)				
Eurasian Hoopoe	*Upapa epops*	戴胜	Summer visitor/passage migrant/ winter visitor	LC
Megalaimidae (Asian Barbets)				
Great Barbet	*Megalaima virens*	大拟啄木鸟	Resident	LC
Green-eared Barbet	*Megalaima faiostricta*	黄纹拟啄木鸟	Resident	LC
Chinese Barbet	*Megalaima faber*	黑眉拟啄木鸟	Resident	LC
Picidae (Woodpeckers)				
Eurasian Wryneck	*Jynx torquilla*	蚁䴕	Winter visitor	LC
Speckled Piculet	*Picumnus innominatus*	斑姬啄木鸟	Resident	LC
White-browed Piculet	*Sasia ochracea*	白眉棕啄木鸟	Resident	LC
Rufous-bellied Woodpecker	*Dendrocopos hyperythrus*	棕腹啄木鸟	Winter visitor	LC
Grey-capped Pygmy Woodpecker	*Dendrocopos canicapillus*	星头啄木鸟	Resident	LC

Common name	Scientific name	Chinese name	Status	IUCN Red List Status
Crimson-breasted Woodpecker	Dendrocopos cathpharius	赤胸啄木鸟	Resident	LC
White-backed Woodpecker	Dendrocopos leucotos	白背啄木鸟	Resident	LC
Great Spotted Woodpecker	Dendrocopos major	大斑啄木鸟	Resident	LC
Greater Yellownape	Chrysophlegma flavinucha	黄冠啄木鸟	Resident	LC
Lesser Yellownape	Picus chlorolophus	大黄冠啄木鸟	Resident	LC
Grey-headed Woodpecker	Picus canus	灰头绿啄木鸟	Resident	LC
Pale-headed Woodpecker	Gecinulus grantia	竹啄木鸟	Resident	LC
Bay Woodpecker	Blythipicus pyrrhotis	黄嘴栗啄木鸟	Resident	LC
Rufous Woodpecker	Micropternus brachyurus	栗啄木鸟	Resident	LC
Eurylaimidae (Broadbills)				
Silver-breasted Broadbill	Serilophus lunatus	银胸丝冠鸟	Resident	LC
Pittidae (Pittas)				
Blue-rumped Pitta	Hydrornis soror	蓝背八色鸫	Resident	LC
Fairy Pitta	Pitta nympha	仙八色鸫	Summer visitor/passage migrant	VU
Blue-winged Pitta	Pitta moluccensis	蓝翅八色鸫	Vagrant	LC
Tephrodornithidae (Woodshrikes and allies)				
Bar-winged Flycatcher-shrike	Hemipus picatus	褐背鹟鵙	Resident	LC
Large Woodshrike	Tephrodornis gularis	钩嘴林鵙	Resident	LC
Artamidae (Woodswallows)				
Ashy Woodswallow	Artamus fuscus	灰燕鵙	Non-breeding visitor	LC
Campephagidae (Cuckooshrikes)				
Large Cuckooshrike	Coracina macei	大鹃鵙	Resident	LC
Black-winged Cuckooshrike	Coracina melanoschistos	暗灰鹃鵙	Summer visitor/winter visitor	LC
Rosy Minivet	Pericrocotus roseus	粉红山椒鸟	Summer visitor	LC
Swinhoe's Minivet	Pericrocotus cantonensis	小灰山椒鸟	Summer visitor	LC
Ashy Minivet	Pericrocotus divaricatus	灰山椒鸟	Passage migrant	LC
Grey-chinned Minivet	Pericrocotus solaris	灰喉山椒鸟	Resident	LC
Long-tailed Minivet	Pericrocotus ethologus	长尾山椒鸟	Summer visitor	LC
Short-billed Minivet	Pericrocotus brevirostris	短嘴山椒鸟	Resident	LC
Scarlet Minivet	Pericrocotus speciosus	赤红山椒鸟	Resident	LC
Laniidae (Shrikes)				
Tiger Shrike	Lanius tigrinus	虎纹伯劳	Passage migrant	LC
Bull-headed Shrike	Lanius bucephalus	牛头伯劳	Winter visitor	LC
Brown Shrike	Lanius cristatus	红尾伯劳	Summer visitor/passage migrant/ winter visitor	LC
Burmese Shrike	Lanius collurioides	栗背伯劳	Vagrant	LC
Long-tailed Shrike	Lanius schach	棕背伯劳	Resident	LC
Grey-backed Shrike	Lanius tephronotus	灰背伯劳	Resident	LC
Chinese Grey Shrike	Lanius sphenocercus	楔尾伯劳	Winter visitor	LC
Vireonidae (Vireos, Greenlets)				
White-bellied Erpornis	Erpornis zantholeuca	白腹凤鹛	Resident	LC
Blyth's Shrike Babbler	Pteruthius aeralatus	红翅鵙鹛	Resident	LC
Green Shrike Babbler	Pteruthius xanthochlorus	淡绿鵙鹛	Resident	LC
Clicking Shrike Babbler	Pteruthius intermedius	栗额鵙鹛	Resident	LC
Oriolidae (Orioles)				
Black-naped Oriole	Oriolus chinensis	黑枕黄鹂	Summer visitor/passage migrant	LC
Maroon Oriole	Oriolus traillii	朱鹂	Resident	LC
Silver Oriole	Oriolus mellianus	鹊色鹂	Summer visitor	EN
Dicruridae (Drongos)				
Black Drongo	Dicrurus macrocercus	黑卷尾	Resident/summer visitor/winter visitor	LC
Ashy Drongo	Dicrurus leucophaeus	灰卷尾	Summer visitor/passage migrant/ winter visitor	LC
Crow-billed Drongo	Dicrurus annectans	鸦嘴卷尾	Summer visitor	LC
Bronzed Drongo	Dicrurus aeneus	古铜色卷尾	Resident	LC
Hair-crested Drongo	Dicrurus hottentotus	发冠卷尾	Resident/summer visitor/passage migrant	LC
Greater Racket-tailed Drongo	Dicrurus paradiseus	大盘尾	Resident	LC
Rhipiduridae (Fantails)				
White-throated Fantail	Rhipidura albicollis	白喉扇尾鹟	Resident	LC
Monarchidae (Monarchs)				
Black-naped Monarch	Hypothymis azurea	黑枕王鹟	Winter visitor/resident/passage migrant	LC
Asian Paradise Flycatcher	Terpsiphone paradisi	寿带	Summer visitor/passage migrant	LC
Japanese Paradise Flycatcher	Terpsiphone atrocaudata	紫寿带	Passage migrant	NT

Common name	Scientific name	Chinese name	Status	IUCN Red List Status
Corvidae (Crows and Jays)				
Eurasian Jay	Garrulus glandarius	松鸦	Resident	LC
Azure-winged Magpie	Cyanopica cyanus	灰喜鹊	Resident	LC
Red-billed Blue Magpie	Urocissa erythroryncha	红嘴蓝鹊	Resident	LC
White-winged Magpie	Urocissa whiteheadi	白翅蓝鹊	Resident	LC
Indochinese Green Magpie	Cissa hypoleuca	印支绿鹊	Resident	LC
Grey Treepie	Dendrocitta formosae	灰树鹊	Resident	LC
Ratchet-tailed Treepie	Temnurus temnurus	塔尾树鹊	Resident	LC
Eurasian Magpie	Pica pica	喜鹊	Resident	LC
Spotted Nutcracker	Nucifraga caryocatactes	星鸦	Resident	LC
Red-billed Chough	Pyrrhocorax pyrrhocorax	红嘴山鸦	Non-breeding visitor	LC
Daurian Jackdaw	Coloeus dauuricus	达乌里寒鸦	Winter visitor	LC
Rook	Corvus frugilegus	秃鼻乌鸦	Winter visitor	LC
Carrion Crow	Corvus corone	小嘴乌鸦	Non-breeding visitor/winter visitor?	LC
Collared Crow	Corvus torquatus	白颈鸦	Resident	NT
Large-billed Crow	Corvus macrorhynchos	大嘴乌鸦	Resident	LC
Bombycillidae (Waxwings)				
Bohemian Waxwing	Bombycilla garrulus	太平鸟	Winter visitor	LC
Japanese Waxwing	Bombycilla japonica	小太平鸟	Winter visitor	NT
Stenostiridae (Fairy Flycatchers)				
Grey-headed Canary Flycatcher	Culicapa ceylonensis	方尾鹟	Summer visitor/winter visitor	LC
Paridae (Tits, Chickadees)				
Yellow-browed Tit	Sylviparus modestus	黄眉林雀	Resident	LC
Sultan Tit	Melanochlora sultanea	冕雀	Resident	LC
Coal Tit	Periparus ater	煤山雀	Resident	LC
Yellow-bellied Tit	Pardaliparus venustulus	黄腹山雀	Resident/non-breeding visitor	LC
Varied Tit	Sittiparus varius	杂色山雀	Resident/non-breeding visitor?	LC
Marsh Tit	Poecile palustris	沼泽山雀	Resident	LC
Japanese Tit	Parus minor	远东山雀	Resident	LC
Cinereous Tit	Parus cinereus	苍背山雀	Resident	LC
Green-backed Tit	Parus monticolus	绿背山雀	Resident	LC
Yellow-cheeked Tit	Machlolophus spilonotus	黄颊山雀	Resident	LC
Remizidae (Penduline Tits)				
Chinese Penduline Tit	Remiz consobrinus	中华攀雀	Winter visitor	LC
Panuridae (Bearded Reedling)				
Bearded Reedling	Panurus biarmicus	文须雀	Vagrant	LC
Alaudidae (Larks)				
Greater Short-toed Lark	Calandrella brachydactyla	大短趾百灵	Vagrant	LC
Asian Short-toed Lark	Calandrella cheleensis	亚洲短趾百灵	Vagrant	LC
Crested Lark	Galerida cristata	凤头百灵	Vagrant	LC
Eurasian Skylark	Alauda arvensis	云雀	Winter visitor	LC
Oriental Skylark	Alauda gulgula	小云雀	Resident/winter visitor	LC
Pycnonotidae (Bulbuls)				
Collared Finchbill	Spizixos semitorques	领雀嘴鹎	Resident	LC
Red-whiskered Bulbul	Pycnonotus jocosus	红耳鹎	Resident	LC
Brown-breasted Bulbul	Pycnonotus xanthorrhous	黄臀鹎	Resident	LC
Light-vented Bulbul	Pycnonotus sinensis	白头鹎	Resident	LC
Sooty-headed Bulbul	Pycnonotus aurigaster	白喉红臀鹎	Resident	LC
Puff-throated Bulbul	Alophoixus pallidus	白喉冠鹎	Resident	LC
Mountain Bulbul	Ixos mcclellandii	绿翅短脚鹎	Resident	LC
Chestnut Bulbul	Hemixos castanonotus	栗背短脚鹎	Resident	LC
Black Bulbul	Hypsipetes leucocephalus	黑短脚鹎	Resident	LC
Brown-eared Bulbul	Hypsipetes amaurotis	栗耳短脚鹎	Winter visitor	LC
Hirundinidae (Swallows and Martins)				
Sand Martin	Riparia riparia	崖沙燕	Winter visitor/passage migrant	LC
Pale Martin	Riparia diluta	淡色沙燕	Passage migrant/winter visitor	LC
Barn Swallow	Hirundo rustica	家燕	Summer visitor/passage migrant/winter visitor	LC
Eurasian Crag Martin	Ptyonoprogne rupestris	岩燕	Vagrant	LC
Common House Martin	Delichon urbica	白腹毛脚燕	Winter visitor/passage migrant	LC
Asian House Martin	Delichon dasypus	烟腹毛脚燕	Summer visitor/passage migrant	LC
Red-rumped Swallow	Cecropis daurica	金腰燕	Summer visitor/passage migrant	LC

Common name	Scientific name	Chinese name	Status	IUCN Red List Status
Pnoepygidae (Wren-Babblers)				
Pygmy Wren-Babbler	Pnoepyga pusilla	小鳞胸鹪鹛	Resident	LC
Cettiidae (Bush Warblers and allies)				
Rufous-faced Warbler	Abroscopus albogularis	棕脸鹟莺	Resident	LC
Mountain Tailorbird	Phyllergates cuculatus	金头缝叶莺	Resident	LC
Japanese Bush Warbler	Horornis diphone	日本树莺	Passage migrant/winter visitor	LC
Manchurian Bush Warbler	Horornis borealis	远东树莺	Winter visitor	LC
Brown-flanked Bush Warbler	Horornis fortipes	强脚树莺	Resident	LC
Yellow-bellied Bush Warbler	Horornis acanthizoides	黄腹树莺	Resident	LC
Grey-bellied Tesia	Tesia cyaniventer	灰腹地莺	Resident/non-breeding visitor	LC
Asian Stubtail	Urosphena squameiceps	鳞头树莺	Winter visitor/passage migrant	LC
Pale-footed Bush Warbler	Urosphena pallidipes	淡脚树莺	Passage migrant	LC
Aegithalidae (Bushtits)				
Silver-throated Bushtit	Aegithalos glaucogularis	银喉长尾山雀	Resident	LC
Black-throated Bushtit	Aegithalos concinnus	红头长尾山雀	Resident	LC
Phylloscopidae (Leaf Warblers and Allies)				
Common Chiffchaff	Phylloscopus collybita	叽喳柳莺	Vagrant	LC
Dusky Warbler	Phylloscopus fuscatus	褐柳莺	Winter visitor/passage migrant	LC
Alpine Leaf Warbler	Phylloscopus occisinensis	华西柳莺	Summer visitor	LC
Buff-throated Warbler	Phylloscopus subaffinis	棕腹柳莺	Summer visitor/winter visitor	LC
Yellow-streaked Warbler	Phylloscopus armandii	棕眉柳莺	Summer visitor/passage migrant	LC
Radde's Warbler	Phylloscopus schwarzi	巨嘴柳莺	Passage migrant	LC
Chinese Leaf Warbler	Phylloscopus yunnanensis	云南柳莺	Passage migrant	LC
Pallas's Leaf Warbler	Phylloscopus proregulus	黄腰柳莺	Winter visitor/passage migrant	LC
Yellow-browed Warbler	Phylloscopus inornatus	黄眉柳莺	Winter visitor/passage migrant	LC
Hume's Leaf Warbler	Phylloscopus humei	淡眉柳莺	Winter visitor/passage migrant	LC
Arctic Warbler	Phylloscopus borealis	极北柳莺	Passage migrant/winter visitor	LC
Kamchatka Leaf Warbler	Phylloscopus examinandus	堪察加柳莺	Passage migrant	LC
Japanese Leaf Warbler	Phylloscopus xanthodryas	日本柳莺	Passage migrant	LC
Two-barred Warbler	Phylloscopus plumbeitarsus	双斑绿柳莺	Winter visitor/passage migrant	LC
Pale-legged Warbler	Phylloscopus tenellipes	淡脚柳莺	Passage migrant	LC
Sakhalin Leaf Warbler	Phylloscopus borealoides	库页岛柳莺	Passage migrant	LC
Large-billed Warbler	Phylloscopus magnirostris	乌嘴柳莺	Summer visitor	LC
Eastern Crowned Warbler	Phylloscopus coronatus	冕柳莺	Passage migrant	LC
Blyth's Leaf Warbler	Phylloscopus reguloides	西南冠纹柳莺	Summer visitor	LC
Claudia's Leaf Warbler	Phylloscopus claudiae	冠纹柳莺	Summer visitor/passage migrant	LC
Hartert's Leaf Warbler	Phylloscopus goodsoni	华南冠纹柳莺	Summer visitor/winter visitor	LC
Emei Leaf Warbler	Phylloscopus emeiensis	峨眉柳莺	Summer visitor	LC
Davison's Leaf Warbler	Phylloscopus davisoni	云南白斑尾柳莺	Resident	LC
Kloss's Leaf Warbler	Phylloscopus ogilviegranti	白斑尾柳莺	Summer visitor	LC
Hainan Leaf Warbler	Phylloscopus hainanus	海南柳莺	Resident	VU
Sulphur-breasted Warbler	Phylloscopus ricketti	黑眉柳莺	Summer visitor	LC
White-spectacled Warbler	Seicercus affinis	白眶鹟莺	Summer visitor/winter visitor	LC
Grey-crowned Warbler	Seicercus tephrocephalus	灰冠鹟莺	Summer visitor	LC
Bianchi's Warbler	Seicercus valentini	比氏鹟莺	Summer visitor/passage migrant	LC
Marten's Warbler	Seicercus omeiensis	峨嵋鹟莺	Summer visitor/passage migrant	LC
Plain-tailed Warbler	Seicercus soror	淡尾鹟莺	Summer visitor/passage migrant	LC
Chestnut-crowned Warbler	Seicercus castaniceps	栗头鹟莺	Summer visitor/winter visitor	LC
Acrocephalidae (Reed Warblers and Allies)				
Oriental Reed Warbler	Acrocephalus orientalis	东方大苇莺	Summer visitor/passage migrant	LC
Black-browed Reed Warbler	Acrocephalus bistrigiceps	黑眉苇莺	Summer visitor/passage migrant	LC
Streaked Reed Warbler	Acrocephalus sorghophilus	细纹苇莺	Passage migrant	EN
Blunt-winged Warbler	Acrocephalus concinens	钝翅苇莺	Summer visitor/passage migrant	LC
Manchurian Reed Warbler	Acrocephalus tangorum	远东苇莺	Passage migrant	VU
Thick-billed Warbler	Iduna aedon	厚嘴苇莺	Passage migrant	LC
Locustellidae (Grassbirds and Allies)				
Russet Bush Warbler	Locustella mandelli	高山短翅莺	Resident/summer visitor/winter visitor	LC
Baikal Bush Warbler	Locustella davidi	北短翅莺	Passage migrant	LC
Chinese Bush Warbler	Locustella tacsanowskia	中华短翅莺	Summer visitor/passage migrant	LC
Brown Bush Warbler	Locustella luteoventris	棕褐短翅莺	Resident	LC
Lanceolated Warbler	Locustella lanceolata	矛斑蝗莺	Passage migrant	LC
Middendorff's Grasshopper Warbler	Locustella ochotensis	北蝗莺	Passage migrant	LC

Common name	Scientific name	Chinese name	Status	IUCN Red List Status
Styan's Grasshopper Warbler	Locustella pleskei	史氏蝗莺	Winter visitor	VU
Pallas's Grasshopper Warbler	Locustella certhiola	小蝗莺	Passage migrant	LC
Marsh Grassbird	Locustella pryeri	斑背大尾莺	Resident/winter visitor	NT
Gray's Grasshopper Warbler	Locustella fasciolata	苍眉蝗莺	Passage migrant	LC
Cisticolidae (Cisticolas and Allies)				
Zitting Cisticola	Cisticola juncidis	棕扇尾莺	Resident	LC
Golden-headed Cisticola	Cisticola exilis	金头扇尾莺	Resident	LC
Striated Prinia	Prinia crinigera	山鹪莺	Resident	LC
Hill Prinia	Prinia superciliaris	黑喉山鹪莺	Resident	LC
Rufescent Prinia	Prinia rufescens	暗冕山鹪莺	Resident	LC
Yellow-bellied Prinia	Prinia flaviventris	黄腹山鹪莺	Resident	LC
Plain Prinia	Prinia inornata	纯色山鹪莺	Resident	LC
Common Tailorbird	Orthotomus sutorius	长尾缝叶莺	Resident	LC
Elachuridae (Elachura)				
Elachura	Elachura formosus	丽星鹩鹛	Resident	
Timaliidae (Timaliid Babblers)				
Large Scimitar Babbler	Pomatorhinus hypoleucos	长嘴钩嘴鹛	Resident	LC
Black-streaked Scimitar Babbler	Pomatorhinus gravivox	斑胸钩嘴鹛	Resident	LC
Grey-sided Scimitar Babbler	Pomatorhinus swinhoei	华南斑胸钩嘴鹛	Resident	LC
Streak-breasted Scimitar Babbler	Pomatorhinus ruficollis	棕颈钩嘴鹛	Resident	LC
Spot-necked Babbler	Stachyris strialata	斑颈穗鹛	Resident	LC
Rufous-capped Babbler	Stachyridopsis ruficeps	红头穗鹛	Resident	LC
Chestnut-capped Babbler	Timalia pileata	红顶鹛	Resident	LC
Pellorneidae (Fulvettas and Ground Babblers)				
Rusty-capped Fulvetta	Alcippe dubia	褐胁雀鹛	Resident	LC
Dusky Fulvetta	Alcippe brunnea	褐顶雀鹛	Resident	LC
David's Fulvetta	Alcippe davidi	灰眶雀鹛	Resident	LC
Huet's Fulvetta	Alcippe hueti	淡眉雀鹛	Resident	LC
Eyebrowed Wren-Babbler	Napothera epilepidota	纹胸鹩鹛	Resident	LC
Chinese Grassbird	Graminicola striatus	大草莺	Resident	NT
Leiothrichidae (Laughingthrushes)				
Chinese Babax	Babax lanceolatus	矛纹草鹛	Resident	LC
Chinese Hwamei	Garrulax canorus	画眉	Resident	LC
Grey Laughingthrush	Garrulax maesi	褐胸噪鹛	Resident	LC
Rufous-cheeked Laughingthrush	Garrulax castanotis	栗颊噪鹛	Resident	LC
Moustached Laughingthrush	Garrulax cineraceus	灰翅噪鹛	Resident	LC
Spotted Laughingthrush	Garrulax ocellatus	眼纹噪鹛	Resident	LC
Masked Laughingthrush	Garrulax perspicillatus	黑脸噪鹛	Resident	LC
White-throated Laughingthrush	Garrulax albogularis	白喉噪鹛	Resident	LC
Lesser Necklaced Laughingthrush	Garrulax monileger	小黑领噪鹛	Resident	LC
Greater Necklaced Laughingthrush	Garrulax pectoralis	黑领噪鹛	Resident	LC
Black-throated Laughingthrush	Garrulax chinensis	黑喉噪鹛	Resident	LC
Swinhoe's Laughingthrush	Garrulax monachus	海南噪鹛	Resident	NE
Blue-crowned Laughingthrush	Garrulax courtoisi	靛冠噪鹛	Resident	CR
Buffy Laughingthrush	Garrulax berthemyi	棕噪鹛	Resident	LC
White-browed Laughingthrush	Garrulax sannio	白颊噪鹛	Resident	LC
Elliot's Laughingthrush	Trochalopteron elliotii	橙翅噪鹛	Resident	LC
Red-winged Laughingthrush	Trochalopteron formosum	丽色噪鹛	Resident	LC
Red-tailed Laughingthrush	Trochalopteron milnei	赤尾噪鹛	Resident	LC
Blue-winged Minla	Minla cyanouroptera	蓝翅希鹛	Resident	LC
Red-tailed Minla	Minla ignotincta	火尾希鹛	Resident	LC
Red-billed Leiothrix	Leiothrix lutea	红嘴相思鸟	Resident	LC
Black-headed Sibia	Heterophasia desgodinsi	黑头奇鹛	Resident	LC
Sylviidae (Sylviid Babblers)				
Golden-breasted Fulvetta	Lioparus chrysotis	金胸雀鹛	Resident	LC
Grey-hooded Fulvetta	Fulvetta cinereiceps	灰头雀鹛	Resident	LC
Yellow-eyed Babbler	Chrysomma sinense	金眼鹛雀	Resident?	LC
Spectacled Parrotbill	Sinosuthora conspicillata	白眶鸦雀	Resident	LC
Vinous-throated Parrotbill	Sinosuthora webbiana	棕头鸦雀	Resident	LC
Golden Parrotbill	Suthora verreauxi	金色鸦雀	Resident	LC
Short-tailed Parrotbill	Neosuthora davidiana	短尾鸦雀	Resident	LC
Grey-headed Parrotbill	Psittiparus gularis	灰头鸦雀	Resident	LC

Common name	Scientific name	Chinese name	Status	IUCN Red List Status
Spot-breasted Parrotbill	*Paradoxornis guttaticollis*	点胸鸦雀	Resident	LC
Reed Parrotbill	*Paradoxornis heudei*	震旦鸦雀	Resident	NT
Zosteropidae (White-eyes)				
Chestnut-collared Yuhina	*Yuhina torqueola*	栗颈凤鹛	Resident	LC
White-collared Yuhina	*Yuhina diademata*	白领凤鹛	Resident	LC
Black-chinned Yuhina	*Yuhina nigrimenta*	黑颏凤鹛	Resident	LC
Chestnut-flanked White-eye	*Zosterops erythropleurus*	红胁绣眼鸟	Passage migrant	LC
Japanese White-eye	*Zosterops palpebrosus*	暗绿绣眼鸟	Resident/winter visitor	LC
Regulidae (Goldcrests, Kinglets)				
Goldcrest	*Regulus regulus*	戴菊	Winter visitor	LC
Troglodytidae (Wrens)				
Eurasian Wren	*Troglodytes troglodytes*	鹪鹩	Winter visitor	LC
Sittidae (Nuthatches)				
Eurasian Nuthatch	*Sitta europaea*	普通鸭	Resident	LC
Chestnut-vented Nuthatch	*Sitta nagaensis*	栗臀鸭	Resident	LC
Velvet-fronted Nuthatch	*Sitta frontalis*	绒额鸭	Resident (introduced)	LC
Yellow-billed Nuthatch	*Sitta solangiae*	淡紫鸭	Resident	NT
Tichodromidae (Wallcreeper)				
Wallcreeper	*Tichodroma muraria*	红翅旋壁雀	Vagrant	LC
Sturnidae (Starlings)				
Hill Myna	*Gracula religiosa*	鹩哥	Resident	LC
Crested Myna	*Acridotheres cristatellus*	八哥	Resident	LC
Common Myna	*Acridotheres tristis*	家八哥	Resident (Introduced)	LC
Red-billed Starling	*Spodiopsar sericeus*	丝光椋鸟	Resident/winter visitor	LC
White-cheeked Starling	*Spodiopsar cineraceus*	灰椋鸟	Winter visitor	LC
Black-collared Starling	*Gracupica nigricollis*	黑领椋鸟	Resident	LC
Asian Pied Starling	*Gracupica contra*	斑椋鸟	Resident (Introduced)	LC
Daurian Starling	*Agropsar sturninus*	北椋鸟	Passage migrant	LC
Chestnut-cheeked Starling	*Agropsar philippensis*	紫背椋鸟	Passage migrant	LC
White-shouldered Starling	*Sturnia sinensis*	灰背椋鸟	Summer visitor	LC
Chestnut-tailed Starling	*Sturnia malabarica*	灰头椋鸟	Vagrant	LC
Rosy Starling	*Pastor roseus*	粉红椋鸟	Vagrant	LC
Common Starling	*Sturnus vulgaris*	紫翅椋鸟	Winter visitor	LC
Turdidae (Thrushes)				
Orange-headed Thrush	*Geokichla citrina*	橙头地鸫	Summer visitor/passage migrant/ winter visitor	LC
Siberian Thrush	*Geokichla sibirica*	白眉地鸫	Passage migrant/winter visitor	LC
White's Thrush	*Zoothera aurea*	怀氏虎鸫	Winter visitor	LC
Plain-backed Thrush	*Zoothera mollissima*	光背地鸫	Winter visitor	LC
Grey-backed Thrush	*Turdus hortulorum*	灰背鸫	Winter visitor	LC
Japanese Thrush	*Turdus cardis*	乌灰鸫	Winter visitor	LC
Grey-winged Blackbird	*Turdus boulboul*	灰翅鸫	Summer visitor	LC
Common Blackbird	*Turdus merula*	乌鸫	Resident/winter visitor	LC
Chestnut Thrush	*Turdus rubrocanus*	灰头鸫	Summer visitor	LC
Grey-sided Thrush	*Turdus feae*	褐头鸫	Vagrant	VU
Eyebrowed Thrush	*Turdus obscurus*	白眉鸫	Passage migrant/winter visitor	LC
Pale Thrush	*Turdus pallidus*	白腹鸫	Passage migrant/winter visitor	LC
Brown-headed Thrush	*Turdus chrysolaus*	赤胸鸫	Passage migrant/winter visitor	LC
Black-throated Thrush	*Turdus atrogularis*	黑颈鸫	Vagrant	LC
Red-throated Thrush	*Turdus ruficollis*	赤颈鸫	Vagrant	LC
Naumann's Thrush	*Turdus naumanni*	红尾鸫	Winter visitor/passage migrant	LC
Dusky Thrush	*Turdus eunomus*	斑鸫	Winter visitor/passage migrant	LC
Chinese Thrush	*Turdus mupinensis*	宝兴歌鸫	Winter visitor	LC
Green Cochoa	*Cochoa viridis*	绿宽嘴鸫	Resident	LC
Muscicapidae (Chats and Old World Flycatchers)				
Japanese Robin	*Erithacus akahige*	日本歌鸲	Winter visitor/passage migrant	LC
Lesser Shortwing	*Brachypteryx leucophris*	白喉短翅鸫	Resident	LC
White-browed Shortwing	*Brachypteryx montana*	蓝短翅鸫	Resident	LC
Bluethroat	*Luscinia svecica*	蓝喉歌鸲	Passage migrant/winter visitor	LC
Siberian Rubythroat	*Luscinia calliope*	红喉歌鸲	Passage migrant/winter visitor	LC
Siberian Blue Robin	*Luscinia cyane*	蓝歌鸲	Passage migrant	LC
Rufous-tailed Robin	*Luscinia sibilans*	红尾歌鸲	Winter visitor/passage migrant	LC
Orange-flanked Bluetail	*Tarsiger cyanurus*	红胁蓝尾鸲	Passage migrant/winter visitor	LC
Oriental Magpie Robin	*Copsychus saularis*	鹊鸲	Resident	LC
White-rumped Shama	*Copsychus malabaricus*	白腰鹊鸲	Resident	LC

Common name	Scientific name	Chinese name	Status	IUCN Red List Status
Common Redstart	*Phoenicurus phoenicurus*	欧亚红尾鸲	Vagrant	LC
Hodgson's Redstart	*Phoenicurus hodgsoni*	黑喉红尾鸲	Non-breeding visitor	LC
Daurian Redstart	*Phoenicurus auroreus*	北红尾鸲	Passage migrant/winter visitor	LC
Blue-fronted Redstart	*Phoenicurus frontalis*	蓝额红尾鸲	Non-breeding visitor	LC
Plumbeous Water Redstart	*Rhyacornis fuliginosa*	红尾水鸲	Resident	LC
White-capped Water Redstart	*Chaimarrornis leucocephalus*	白顶溪鸲	Resident	LC
White-tailed Robin	*Myiomela leucura*	白尾蓝地鸲	Resident/summer visitor	LC
Blue Whistling Thrush	*Myophonus caeruleus*	紫啸鸫	Resident	LC
Little Forktail	*Enicurus scouleri*	小燕尾	Resident	LC
Slaty-backed Forktail	*Enicurus schistaceus*	灰背燕尾	Resident	LC
White-crowned Forktail	*Enicurus leschenaulti*	白冠燕尾	Resident	LC
Spotted Forktail	*Enicurus maculatus*	斑背燕尾	Resident	LC
Siberian Stonechat	*Saxicola maurus*	黑喉石即鸟	Passage migrant/winter visitor	LC
Grey Bushchat	*Saxicola ferreus*	灰林即鸟	Summer visitor/winter visitor	LC
Isabelline Wheatear	*Oenanthe isabellina*	沙即鸟	Vagrant	LC
Northern Wheatear	*Oenanthe oenanthe*	穗即鸟	Vagrant	LC
Blue Rock Thrush	*Monticola solitarius*	蓝矶鸫	Resident/summer visitor/winter visitor	LC
Chestnut-bellied Rock Thrush	*Monticola rufiventris*	栗腹矶鸫	Resident/winter visitor	LC
White-throated Rock Thrush	*Monticola gularis*	白喉矶鸫	Passage migrant/winter visitor	LC
Brown-chested Jungle Flycatcher	*Rhinomyias brunneatus*	白喉林鹟	Summer visitor/passage migrant	VU
Grey-streaked Flycatcher	*Muscicapa griseisticta*	灰纹鹟	Passage migrant	LC
Dark-sided Flycatcher	*Muscicapa sibirica*	乌鹟	Passage migrant	LC
Asian Brown Flycatcher	*Muscicapa latirostris*	北灰鹟	Passage migrant/winter visitor	LC
Brown-breasted Flycatcher	*Muscicapa muttui*	褐胸鹟	Summer visitor	LC
Ferruginous Flycatcher	*Muscicapa ferruginea*	棕尾褐鹟	Passage migrant/winter visitor	LC
Yellow-rumped Flycatcher	*Ficedula zanthopygia*	白眉姬鹟	Passage migrant	LC
Narcissus Flycatcher	*Ficedula narcissina*	黄眉姬鹟	Passage migrant	LC
Green-backed Flycatcher	*Ficedula elisae*	绿背姬鹟	Passage migrant	LC
Mugimaki Flycatcher	*Ficedula mugimaki*	鸲姬鹟	Passage migrant	LC
Rufous-gorgeted Flycatcher	*Ficedula strophiata*	橙胸姬鹟	Summer visitor/winter visitor	LC
Taiga Flycatcher	*Ficedula albicilla*	红喉姬鹟	Passage migrant/winter visitor	LC
Snowy-browed Flycatcher	*Ficedula hyperythra*	棕胸蓝姬鹟	Resident	LC
Slaty-blue Flycatcher	*Ficedula tricolor*	灰蓝姬鹟	Summer visitor	LC
Blue-and-white Flycatcher	*Cyanoptila cyanomelana*	白腹蓝鹟	Passage migrant	LC
Zappey's Flycatcher	*Cyanoptila cumatilis*	琉璃蓝鹟	Passage migrant	LC
Verditer Flycatcher	*Eumyias thalassinus*	铜蓝鹟	Winter visitor/summer visitor	LC
Hainan Blue Flycatcher	*Cyornis hainanus*	海南蓝仙鹟	Resident/summer visitor	LC
Pale Blue Flycatcher	*Cyornis unicolor*	纯蓝仙鹟	Resident/summer visitor	LC
Chinese Blue Flycatcher	*Cyornis glaucicomans*	中华仙鹟	Summer visitor/winter visitor	LC
Fujian Niltava	*Niltava davidi*	棕腹大仙鹟	Summer visitor/winter visitor	LC
Rufous-bellied Niltava	*Niltava sundara*	棕腹仙鹟	Resident	LC
Small Niltava	*Niltava macgrigoriae*	小仙鹟	Resident	LC
Dippers (Cinclidae)				
Brown Dipper	*Cinclus pallasii*	褐河乌	Resident	LC
Chloropseidae (Leafbirds)				
Orange-bellied Leafbird	*Chloropsis hardwickii*	橙腹叶鹎	Resident	LC
Dicaeidae (Flowerpeckers)				
Plain Flowerpecker	*Dicaeum minullum*	纯色啄花鸟	Resident	LC
Fire-breasted Flowerpecker	*Dicaeum ignipectus*	红胸啄花鸟	Resident	LC
Scarlet-backed Flowerpecker	*Dicaeum cruentatum*	朱背啄花鸟	Resident	LC
Nectariniidae (Sunbirds)				
Olive-backed Sunbird	*Cinnyris jugularis*	黄腹花蜜鸟	Resident	LC
Mrs. Gould's Sunbird	*Aethopyga gouldiae*	蓝喉太阳鸟	Summer visitor/winter visitor	LC
Fork-tailed Sunbird	*Aethopyga christinae*	叉尾太阳鸟	Resident	LC
Crimson Sunbird	*Aethopyga siparaja*	黄腰太阳鸟	Resident	LC
Black-throated Sunbird	*Aethopyga saturata*	黑胸太阳鸟	Resident	LC
Passeridae (Old World Sparrows)				
House Sparrow	*Passer domesticus*	家麻雀	Non-breeding visitor	LC
Russet Sparrow	*Passer rutilans*	山麻雀	Resident/winter visitor	LC
Eurasian Tree Sparrow	*Passer montanus*	麻雀	Resident	LC
Estrildidae (Waxbills, Munias and allies)				
Red Avadavat	*Amandava amandava*	红梅花雀	Resident (Introduced)	LC

Common name	Scientific name	Chinese name	Status	IUCN Red List Status
White-rumped Munia	*Lonchura striata*	白腰文鸟	Resident	LC
Scaly-breasted Munia	*Lonchura punctulata*	斑文鸟	Resident	LC
Chestnut Munia	*Lonchura atricapilla*	栗腹文鸟	Resident (Introduced)	LC
Java Sparrow	*Padda oryzivora*	爪哇禾雀	Resident (Introduced)	VU
Prunellidae (Accentors)				
Rufous-breasted Accentor	*Prunella strophiata*	棕胸岩鹨	Winter visitor	LC
Siberian Accentor	*Prunella montanella*	棕眉山岩鹨	Winter visitor	LC
Motacillidae (Wagtails and Pipits)				
Forest Wagtail	*Dendronanthus indicus*	山鹡鸰	Summer visitor/passage migrant/winter visitor	LC
Eastern Yellow Wagtail	*Motacilla tschutschensis*	黄鹡鸰	Winter visitor/passage migrant	LC
Citrine Wagtail	*Motacilla citreola*	黄头鹡鸰	Passage migrant/winter visitor	LC
Grey Wagtail	*Motacilla cinerea*	灰鹡鸰	Passage migrant/winter visitor	LC
White Wagtail	*Motacilla alba*	白鹡鸰	Summer visitor/passage migrant/winter visitor	LC
Richard's Pipit	*Anthus richardi*	理氏鹨	Resident/winter visitor	LC
Meadow Pipit	*Anthus pratensis*	草地鹨	Vagrant	LC
Tree Pipit	*Anthus trivialis*	林鹨	Vagrant	LC
Olive-backed Pipit	*Anthus hodgsoni*	树鹨	Passage migrant/winter visitor	LC
Pechora Pipit	*Anthus gustavi*	北鹨	Passage migrant	LC
Rosy Pipit	*Anthus roseatus*	粉红胸鹨	Summer visitor/passage migrant/winter visitor	LC
Red-throated Pipit	*Anthus cervinus*	红喉鹨	Winter visitor	LC
Buff-bellied Pipit	*Anthus rubescens*	黄腹鹨	Winter visitor	LC
Water Pipit	*Anthus spinoletta*	水鹨	Winter visitor	LC
Upland Pipit	*Anthus sylvanus*	山鹨	Resident	LC
Fringillidae (Finches)				
Brambling	*Fringilla montifringilla*	燕雀	Winter visitor	LC
Hawfinch	*Coccothraustes coccothraustes*	锡嘴雀	Winter visitor	LC
Chinese Grosbeak	*Eophona migratoria*	黑尾蜡嘴雀	Winter visitor/passage migrant/summer visitor	LC
Japanese Grosbeak	*Eophona personata*	黑头蜡嘴雀	Passage migrant/winter visitor	LC
Brown Bullfinch	*Pyrrhula nipalensis*	褐灰雀	Resident	LC
Eurasian Bullfinch	*Pyrrhula pyrrhula*	红腹灰雀	Winter visitor	LC
Common Rosefinch	*Carpodacus erythrinus*	普通朱雀	Winter visitor	LC
Vinaceous Rosefinch	*Carpodacus vinaceus*	酒红朱雀	Resident/non-breeding visitor	LC
Pallas's Rosefinch	*Carpodacus roseus*	北朱雀	Winter visitor	LC
Grey-capped Greenfinch	*Chloris sinica*	金翅雀	Resident/winter visitor	LC
Common Redpoll	*Acanthis flammea*	白腰朱顶雀	Vagrant	LC
Red Crossbill	*Loxia curvirostra*	红交嘴雀	Winter visitor	LC
Eurasian Siskin	*Spinus spinus*	黄雀	Winter visitor	LC
Emberizidae (Buntings)				
Crested Bunting	*Emberiza lathami*	凤头鹀	Resident	LC
Slaty Bunting	*Emberiza siemsseni*	蓝鹀	Summer visitor/winter visitor	LC
Pine Bunting	*Emberiza leucocephalos*	白头鹀	Vagrant	LC
Godlewski's Bunting	*Emberiza godlewskii*	戈氏岩鹀	Non-breeding visitor	LC
Meadow Bunting	*Emberiza cioides*	三道眉草鹀	Resident/winter visitor	LC
Tristram's Bunting	*Emberiza tristrami*	白眉鹀	Winter visitor/passage migrant	LC
Chestnut-eared Bunting	*Emberiza fucata*	栗耳鹀	Winter visitor/passage migrant	LC
Little Bunting	*Emberiza pusilla*	小鹀	Winter visitor	LC
Yellow-browed Bunting	*Emberiza chrysophrys*	黄眉鹀	Winter visitor/passage migrant	LC
Rustic Bunting	*Emberiza rustica*	田鹀	Winter visitor	LC
Yellow-throated Bunting	*Emberiza elegans*	黄喉鹀	Winter visitor	LC
Yellow-breasted Bunting	*Emberiza aureola*	黄胸鹀	Winter visitor/passage migrant	EN
Chestnut Bunting	*Emberiza rutila*	栗鹀	Winter visitor/passage migrant	LC
Black-headed Bunting	*Emberiza melanocephala*	黑头鹀	Vagrant	LC
Red-headed Bunting	*Emberiza bruniceps*	褐头鹀	Vagrant	LC
Yellow Bunting	*Emberiza sulphurata*	硫磺鹀	Passage migrant	VU
Black-faced Bunting	*Emberiza spodocephala*	灰头鹀	Winter visitor	LC
Pallas's Bunting	*Emberiza pallasi*	苇鹀	Winter visitor	LC
Ochre-rumped Bunting	*Emberiza yessoensis*	红颈苇鹀	Winter visitor	NT
Reed Bunting	*Emberiza schoeniclus*	芦鹀	Winter visitor	LC
Calcariidae (Longspurs)				
Lapland Longspur	*Calcarius lapponicus*	铁爪鹀	Winter visitor	LC

Many websites, databases and online groups now exist for the birdwatcher to source material on bird images, records, sounds and practical information for travel. There are also a number of birdwatching societies that organize trips and activities. We list some of the most useful ones here, and a number of websites relevant to China in general.

GENERAL INFORMATION

China Bird Watching Network - www.chinabirdnet.org
A very useful site that contains material on birdwatching contacts, education and bird-relevant conservation updates in the region.

BIRDWATCHING SOCIETIES

Wild Bird Society of Shanghai
Website: http://www.shwbs.org
E-mail: shanghaiwbs@gmail.com

Zhejiang Wild Bird Society
Address: Zhejiang Museum of Natural History,
No. 6 Westlake Cultural Square,
Chaohui Street, Hangzhou
(Postal Code: 310014)
Website: www.zjbird.cn
E-mail: fanmonkey@126.com

Fujian Bird Watching Society
Postal address: PO Box No. 01A-226, Fuzhou,
Fujian (Postal code: 350001)
FBWS: www.fjbirds.org
E-mail: fjbirds@126.com

Xiamen Bird Watching Society
Address: Yifu Secondary School, No. 209,
Middle Honglian Road, Xiamen, Fujian
Website: www.xmbirds.org
E-mail: aiffel@hotmail.com

Shenzhen Bird Watching Society
Address: Room 102 Complex Building,
Futian National Nature Reserve Futian,
Shenzhen, Guangdong (Postal Code: 518040)
Website: www.szbird.org.cn
Email: admin@szbird.org.cn

TRIP REPORTS

Cloud Birder – www.cloudbirders.com
The largest repository of birdwatching trip reports on the internet. Trip reports are organised by geographic location and can be sorted by year and month. At the time of writing, there are over 300 reports from China, many of which cover the Southeast.

RECORD DATABASES

China Bird Report (中国观鸟记录中心) **– www.birdtalker.net**
Thousands of bird lists, reports, records and photographs from across China are deposited onto this bilingual online database annually. Records can be sorted by site and time. An excellent source of material to know what species has been recently seen at which sites. This is also the main site for submissions of bird records and lists in China.

Ebird – www.ebird.org
Part of an extensive citizen science project initiated by Cornell University, this database is useful as a tool to submit bird records, as well as check recent bird sightings from sites across the world. Data from China is still limited, but growing steadily with deposits by travelling birdwatchers.

BIRD CHECKLISTS

China Bird Report (CBR) Checklist – www.sites.google.com/site/cbrchinabirdlist/
The most authoritative and updated checklist of the birds of China. All names are in Chinese, English and Latin. Information on threat status of each species is also provided.

IMAGE DATABASES

China Wild Bird Gallery (中国野鸟图库) **– www.cnbirds.org.cn**
An excellent collection of images of almost all species in the China and Taiwan lists, with some contributions from Southeast Asia. Site is in Chinese, with English and Latin names.

Oriental Bird Images - www.orientalbirdimages.org
A large online repository of bird photographs covering the over 2600 species of Asia. A good source of photographic material for difficult-to-identify species as it contains many images for most species, including different races, age and sex.

SOUND DATABASES

Xeno-canto – www.xeno-canto.org
A comprehensive database of bird sounds that can be accessed freely. A good source of sound material to familiarise with the calls of species before heading off for a birdwatching trip. Most sound recordings collected by the authors from fieldwork during the preparation of this book are deposited here.

Macaulay Library of Natural Sounds – www.macaulaylibrary.org
A very comprehensive library of wildlife sound recordings and videos. Most material can be streamed from the website, while sounds can also be purchased with reasonable costs.

OTHER TRAVEL LOGISTICS

Ctrip – www.ctrip.com
A useful website for booking a wide variety of accommodation and flights within China, especially for independent travellers.

RECOMMENDED READING

Books

Brazil, M. (2009) *Birds of East Asia.* Princeton University Press, New Jersey, USA.

Corlett, R.T. (2009) *The Ecology of Tropical East Asia.* Oxford University Press, Oxford, UK.

Chan, S., Crosby, M., So, S., & Hua, F. (2009) Directory of Important Bird Areas in China (Mainland). Birdlife International, Cambridge, UK.

Cheng, S-L., Liu, J-N. & Zhang, Y-Y. (2011) *The Birds of Wuyishan National Nature Reserve.* Science Publishing, Beijing, China. (In Chinese, with English names)

De Schauensee, R.M. (1984) *The Birds of China.* Smithsonian Institution Press, Washington DC, USA.

Lei, F. & Lu, T. (2004) *China Endemic Birds.* Science Publishing, Beijing, China (In Chinese with English names)

Li, Z., Jiang, H. & Lu, Y. (2008) *A Complete Taxonomic Checklist and Geographic Reference of Bird Species and Subspecies in Eastern China.* Science Publishing, Beijing. (In Chinese, with English names)

Li, Z. (2013) *Study on Biodiversity of Forest Birds in the Nature Reserve of Wuyuan, Jiangxi.* Science Press, Beijing. (In Chinese with English names)

Qu, L-M. eds. (2013) *Field Guide to the Birds of China. Volume 1-3.* The Straits Publishing and Distribution Group, China. (In Chinese with English names)

MacKinnon, J. & Phillipps, K. (2000) *A Field Guide to the Birds of China.* Oxford University Press, Oxford, UK.

Robson, C. (2009) *A Field Guide to the Birds of Southeast Asia.* New Holland Publishers Ltd, London.

The Hong Kong Bird Watching Society (2012) *A Photographic Guide to the Birds of Hong Kong* (Revised Edition). Wan Li Book Co Ltd, Hong Kong.

Viney, C., Phillipps, K. & Lam,C.Y. (1996) *Birds of Hong Kong and South China.* 7th Edition. Government Printer, Hong Kong.

Woodward, T. (2011) *Birding South-east China.* Worldwide Fund for Nature and Hong Kong Bird Watching Society, Hong Kong.

Academic Papers

Gao, Y. (1999) Conservation status of endemic Galliformes on Hainan Island, China. *Bird Conservation International* 9: 411–416.

Ge, Z., Wang, T., Zhou, X. & Shi, W. (2006) Seasonal change and habitat selection of shorebird community at the South Yangtze River Mouth and North Hangzhou Bay, China. *Acta Ecologica Sinica* 26: 40-47.

Kwok, H.K. & Corlett, R.C. (2000) The bird communities of a natural secondary forest and a *Lophostemon confertus* plantation in Hong Kong, South China. *Forest Ecology and Management* 130: 227-234.

Lee, K.S., Lau, M.W-N., Fellowes, J.R. & Chan, B.P-L. (2006) Forest bird fauna of South China: notes on current distribution and status. *Forktail* 22: 23-38.

Lewthwaite R.W. (1996) Forest Birds of South-east China: observation during 1984-96. *Hong Kong Bird Report 1995*: 150-203.

Zhang, Q., Han, R. & Zou, F. (2011) Effects of artificial afforestation and successional stage on a lowland forest bird community in southern China. *Forest Ecology and Management* 261: 1738-1749.

Zhou, F. (1987) Guild structure of the forest bird community in Dinghushan. *Acta Ecologica Sinica* 7: 176-184.

Zou, F. & Chen, G. (2003) A study of understory bird communities in tropical mountain rain forest of Jianfengling, Hainan Island, China. *Acta Ecologica Sinica* 24: 510-516.

Zou, F., Yang, Q., Dahmer, T., Cai, J. & Zhang, W. (2006) Habitat use of waterbirds in coastal wetland on Leizhou Peninsula, China. *Waterbirds* 29: 459-464.

Yong, D.L., Liu, Y., Low, B.W., Espanola, C.P., Choi, C-Y. & Kawakami, K. (2014) Migratory songbirds in the East Asian-Australasian Flyway: a review from a conservation perspective. *Bird Conservation International* (In press)

Acknowledgements

The contributions of many people made this book possible. Firstly, we thank our publisher John Beaufoy for bringing our proposal for a photographic guide for Southeast China to reality. We are also grateful to Rosemary Wilkinson, Krystyna Mayer, Bikram Grewal and Gulmohur Design for their immense help in the compilation and editing of the manuscript. We thank Richard Corlett, Huang Qin, Mike Kilburn, Lei Jinyu, Wang Xuejin, Zhao Jian and especially Richard Lewthwaite for their many helpful comments on the manuscript and the checklist. A number of our photographer-friends generously put many of their best images at our disposal. They include: Christian Artuso, Mikael Bauer, Abdelhamid Bizid, Ah Kei Looking@Nature, Sam Chan, Chen Qing, Cheng Heng Yee, Frankie Cheong, Chung Weng Kin, James Eaton, Forrest Fong, Sundev Gombobaatar, Martin Hale, John and Jemi Holmes, Hao Xianing, Kinni Ho, Mike Kilburn, Koel Ko, Le Manh Hung, Lee Kam Cheong, Lee Tiah Khee, Lee Shunda, Lee Yat Ming, Jennifer Leung, Wich'yanan Limparungpatthanakij, Low Choon How, Lu Gang, Daisy O'Neill, Björn Olsen, Stuart Price, Pu Ying, Sun Xiaoming, Tan Gim Cheong, Thiti Tan, Tang Jun, Myron Tay, Ivan Tse, Wallace Tse, Tong Menxiu, Wang Jiyi, Jason Wong Wai Hang, Michelle and Peter Wong, Francis Yap, Zhang Ming, Zhang Yong and Zhao Jian. Finally, we thank our families for their support throughout this project.

▪ INDEX ▪